Probability and Evidence

Probability and Evidence

PAUL HORWICH

CAMBRIDGE
UNIVERSITY PRESS

Shaftesbury Road, Cambridge CB2 8EA, United Kingdom

One Liberty Plaza, 20th Floor, New York, NY 10006, USA

477 Williamstown Road, Port Melbourne, VIC 3207, Australia

314–321, 3rd Floor, Plot 3, Splendor Forum, Jasola District Centre, New Delhi – 110025, India

103 Penang Road, #05–06/07, Visioncrest Commercial, Singapore 238467

Cambridge University Press is part of Cambridge University Press & Assessment, a department of the University of Cambridge.

We share the University's mission to contribute to society through the pursuit of education, learning and research at the highest international levels of excellence.

www.cambridge.org
Information on this title: www.cambridge.org/9781316507018

First published 1982
First paperback edition 2011
Cambridge Philosophy Classics edition 2016

A catalogue record for this publication is available from the British Library

Library of Congress Cataloging-in-Publication data
Names: Horwich, Paul, author.
Title: Probability and evidence / Paul Horwich.
Description: New York : Cambridge University Press, 2016. | Originally published: New York : Cambridge University Press, 1982. | Includes bibliographical references and index.
Identifiers: LCCN 2015049717 | ISBN 9781107142107 (Hardback : alk. paper) | ISBN 9781316507018 (Paperback : alk. paper)
Subjects: LCSH: Science–Philosophy. | Probabilities. | Evidence.
Classification: LCC Q175 .H797 2016 | DDC 501–dc23 LC record available at https://lccn.loc.gov/2015049717

ISBN   978-1-107-14210-7   Hardback
ISBN   978-1-316-50701-8   Paperback

To my mother and father

Contents

Preface to this edition

COLIN HOWSON

While it is difficult to say in general terms what makes a book a classic, it is often not difficult to decide that some given book is one, and to my mind there is no doubt that this book is. Despite the author's disclaimer in his Introduction that he is following a Wittgensteinian strategy in eschewing any grand philosophical scheme, there is one in this book: the Bayesian theory, which Paul Horwich regards as illuminating an entire spectrum of philosophical problems associated with that peculiarly scientific subdomain of epistemology, the so-called scientific method. And illuminate it it certainly does in Horwich's book, a masterpiece of simplicity and elegance that belies the often-challenging nature of the material. Here is one characteristic of a classic that we can tick without further ado.

Probability and Evidence was published at a time when it was becoming increasingly fashionable to deny that the methods of science, allegedly heavily constrained by the socio-economic-cultural milieu in which scientists operate, are in any way *sui generis*, or uniquely meriting a patent of objectivity. Curiously enough, the Bayesian framework for evaluating theories and evidence also subscribes to the view that there is nothing specifically to do with the sciences in the inferential methods it endorses. On the other hand it cedes nothing in the way of objectivity, being nothing less than a *logic* of uncertain inference, an account of how factual hypotheses *in general* are evaluated against evidence, whatever the specific context. This combination of universality and logical rigour was repeatedly stressed by the great twentieth-century Italian Bayesian, Bruno De Finetti, to whom so much of the modern development of the theory is due. But despite its generality the Bayesian theory was developed, by De Finetti and others, primarily to illuminate the procedures of *scientific* inference. As indeed it does, and in this book Horwich shows, with his customary clarity, how it can explain such perennial features of scientific method as why, for example, simpler theories are typically preferred to more complex ones, why diverse evidence is valued more highly than evidence from the same source, why genuinely predictive theories are esteemed more highly than those adjusted to the data, and so on.

Probability and Evidence was itself a pioneering work: it brought what was formerly a sophisticated mathematical theory largely confined to research journals down, if not to the masses, at least to within the reach of a general philosophical readership, or for that matter anyone prepared to do a little work with some relatively simple formulas. Not only is all this very deftly done but – and here is another characteristic of the true classic – several of Horwich's own often highly novel solutions to the problems and paradoxes that figure in the philosophy of science literature became a prominent focus of that literature and – final confirmation of classic status – fixtures of the textbook literature. If someone you know tells you that the notorious raven paradox has been solved, it will most probably be Horwich's solution that they will have in mind.

Preface

My purpose in writing this essay was to exhibit a unified approach to philosophy of science, based on the concept of subjective probability. I hope to have contributed to the subject, first, by offering new treatments of several problems (for example the raven paradox, the nature of surprising data, and the supposed special evidential value of prediction over and above the accommodation of experimental results); and, second, by providing a more complete probabilistic account of scientific methods and assumptions than has been given before. In the interest of autonomy, I have included a chapter on probability which surveys the ideas that will be needed later, keeping technicalities to a minimum. Unfortunately, there was no room for a proper consideration of many alternative points of view; and a much longer treatise would be desirable in which the competition is adequately presented and criticised. Nevertheless, I do make something of a case for the probabilistic approach: it yields satisfying solutions to a wide range of problems in the philosophy of science.

This book is aimed at professional philosophers, but not exclusively; for I have covered a lot of ground, trying to explain everything from scratch, and so I hope that students will find here a useful introduction to the subject. Let no one be intimidated by the occasional intrusion of symbols. The formalism is intended to promote clarity; it is not difficult to master; and if worse comes to worst, those few sections may be skimmed without losing the general drift.

I would like to acknowledge the help and pleasure I have received from Hempel's *Philosophy of Natural Science*, from the work of Hacking and Kyburg on probability, and from Janina Hosiasson-Lindenbaum's brilliant early essay, 'On Confirmation' Also, I am very grateful to Frank Jackson, Thomas Kuhn and Dan Osherson who read an early draft and contributed many helpful suggestions. And I would especially like to thank my friends, Ned Block and Josh Cohen, for their excellent advice and steady encouragement.

Preface

1

Methodology

Introduction

This book is about scientific knowledge, particularly the concept of
evidence. Its purpose is to explore scientific methodology in light of the
obvious yet frequently neglected fact that belief is not an all-or-nothing
matter, but is susceptible to varying degrees of intensity. More specifically,
my main object is to exploit this fact to treat certain well-known puzzles in
the philosophy of science, such as the problem of induction and the
paradox of confirmation, as well as questions about *ad hoc* postulates,
the tenability of realism, statistical testing, the relative merits of prediction
and accommodation, a special quality of varied data, and the evidential
value of further information. My second aim is to display the extent to
which diverse elements of scientific method may be unified and justified
by means of the concept of subjective probability. These two projects are
intimately related. The probabilistic terms in which our evidential ideas
will be formulated should promote clarity and accuracy, dispel confusion,
and thereby facilitate the primary task. I should stress that this main goal
is not to propound or defend a *theory* of the scientific method, either
normative or descriptive, but rather to solve various paradoxical problems. I cannot now adequately describe the conception of philosophy
which promotes this distinction and motivates my approach; but I think
that some appreciation of that metaphilosophical perspective is needed to
understand properly what is being attempted here and to pre-empt certain
objections. I hope the following sketch will be better than nothing.

In a way, philosophy contains science and art. For there are philosophical research programmes whose methodology is scientific and others in
which aesthetic standards prevail. Investigations into the semantics of
natural languages, systematisations of basic ethical judgements, concep-
tual analyses – attempts to formulate necessary and sufficient conditions
for S knows p, x causes y, and w refers to z – these typify scientific
philosophy. Their object is a justified account of certain phenomena – a

simple theory designed to accommodate the relevant data provided, usually, by intuition. On the other wing, we encounter the construction of metaphysical systems – symphonic flights, so far removed from testability that even the rubric of speculation would seem to distort their cognitive status. However, scientific and aesthetic philosophy do not exhaust the subject. What remain are the traditional puzzles and paradoxes, and an accumulating collection of modern ones: the problems of free-will, induction and scepticism, the paradoxes of Epimenides and Newcomb. Typically, we are confronted with apparently good reasons to accept both p and not-p: we have somehow blundered, become attached to some tempting misconception which must be located and exorcised. In such cases, the problem is not to formulate and defend a theory that will dictate which of the contradictory propositions is true; but rather to discover in ourselves the sources of our conflicting inclinations. This is Wittgenstein's idea of pure philosophy: no theories, only 'assembling reminders for a particular purpose': 'a battle against the bewitchment of our intelligence by means of language'. It should be emphasised that from this point of view there is no general reason to impugn the legitimacy or value of whatever else is done in the name of philosophy, and no need to fret about how that word should be used. True, there are bad projects – those resting upon mistaken presuppositions or governed by a foggy collection of adequacy conditions. These, through the confusion they produce, may engender the material for pure philosophy. Nonetheless, any theoretical project may be coherent, provided that there is a definite understanding of how successful accomplishment is to be recognised. And its results can sometimes illuminate a troublesome concept and constitute effective 'reminders' in the struggle to maintain an accurate view of it.

This sort of interaction is exemplified in what follows. The element of what I have called scientific philosophy will consist in the precise characterisation of a notion of degree of belief (designed to have the fruitful property that rational degrees of belief must conform to the probability calculus), and in the explication, within the framework of subjective probability, of various methodological concepts such as 'confirmation', 'ad hoc postulate' and 'diverse evidence'. In order to deflect certain objections, the point of these constructions should be kept in mind. I do not claim that the foundations of subjective probability are absolutely secure, nor that my way is the only or the best way to capture our common-sense idea of degree of confidence. It may be that for other purposes such as psychology, decision theory, the history of science, or even an accurate description of scientific practice, we should prefer an explication which does not commit degrees of belief to precise numerical values. Different accounts of that notion could be appropriate and different analyses of the other relevant concepts may be called for. What is required here is just that the explications be sufficiently faithful and simple to enable a clear perception of the problems under discussion and to permit the confusion and misconceptions which produced them to be exposed and removed.

Aspects of the scientific method

The problems under consideration here stem from a number of very general and widely shared intuitions about evidence, which derive in turn from reflection upon scientific practice. I have divided these intuitions and problems into twelve topics, and what follows is a preliminary discussion of them. They comprise our subject matter; in later chapters each one is treated in more detail, and some explanations and answers are advanced.

(1) *Accommodation of data.* We are inclined to believe theories which make accurate predictions and accommodate experimental results, and disbelieve theories whose consequences are incompatible with our data. This should not require illustration. However, it is worth emphasising the well-known asymmetry between verification and falsification. If a theory is known to entail something false, it is conclusively refuted; but if it is known to entail something true the theory is *not* thereby taken to be conclusively confirmed, but, at best, merely supported. In other words, its correct predictions provide some evidence in favour of a theory, but do not establish it; whereas any mistaken predictions indicate that the theory should be abandoned or at least revised in part.

(2) *Statistical evidence.* The claim, about some experiment, that the probability of a certain outcome is x, is taken to be supported if roughly xn such outcomes, in a long sequence of n instances of the experiment, are obtained; and disconfirmed if the proportion of instances with that outcome differs substantially from x. In particular, if 100 consecutive tosses land heads up we begin to doubt that the coin is fair. In such cases the observed facts are neither entailed nor absolutely precluded by the hypothesis in question. But they are nonetheless of great evidential significance, and it would be desirable to have some explicit rationale for our practice and intuitions regarding the confirmation of empirical probabilistic hypotheses.

(3) *Severe tests.* Theories are tested, and confirmed to some extent if they pass; but they are well confirmed only if the tests are *severe*. For

example, the use of highly accurate measuring instruments will tend to promote the severity of an experimental test, making it more difficult for the theory to pass and more impressive if it does. However, it remains to be seen what is meant in general by 'a severe test', and why survival through such things should give a special boost to the credibility of a theory.

(4) *Surprising predictions.* Particularly powerful support for a theory is conveyed by the verification of its relatively surprising predictions. In other words, a theory gets a lot of credit for predicting something quite unsuspected, or for explaining a bizarre and anomalous phenomenon; and it derives relatively little support from the prediction of something that we expected to occur anyway. So, for example, Einstein's special theory of relativity predicts that clocks moving with high velocity, near the speed of light, will run slowly compared with clocks at rest in our frame of reference, and that this so-called 'time dilation' is detectable. It also predicts that although the same effect is manifested by slowly moving clocks, its magnitude is too insubstantial to be measured, and so slowly moving clocks will *seem* to run at the same rate as stationary clocks. Now, both of these predictions – apparent time dilation in fast, but not in slowly moving clocks – have been verified; yet only the former is taken to provide us with striking confirmation of theory. Why is this? What is it to be surprising? And why is it that surprising, accurate predictions are of special evidential value?

(5) *Paradox of confirmation.* We feel that the hypothesis 'All ravens are black' is significantly confirmed by the observation that certain ravens are black; and not significantly confirmed by the observation that certain things which are not black are also not ravens. But it is peculiarly hard to come up with any rationale for this intuition; and for others like it. The general problem here is known as 'the paradox of confirmation'. It is natural to suppose that any scientific hypothesis of the form 'All As are B' would be supported by evidence of the form 'k is an A and k is a B'. This could well be the sort of principle we might propose as an example of a general canon of scientific methodology. In addition, it seems clear that if some datum supports, or confirms, or is evidence in favour of a scientific hypothesis, then it confirms every logically equivalent formulation of that hypothesis. The trouble – or paradox – is that these two very plausible principles lead to an extremely counterintuitive conclusion. For the first principle tells us that observation of a nonblack nonraven (say a white handkerchief) should be evidence in favour of the hypothesis 'All nonblack things are nonravens'. Therefore, by the second principle, we are driven to the strange conclusion that the hypothesis 'All ravens are black' is confirmed by observation of a white handkerchief. This may be welcome on a rainy day, but it hardly squares with our intuitions about scientific methodology. I will try to explain and justify our intuitions, and show what is wrong with those plausible sounding principles which appear to be incompatible with them.

(6) *The 'grue' problem.* There is a further objection to the natural idea, already threatened by the above paradox of confirmation, that scientific reasoning may be codified by some such rule as

All sampled A s have been B

$\therefore$ Probably, all A s are B

Nelson Goodman (1955) has devised instances of this schema that constitute intuitively bad arguments. For example, define the predicate 'grue' as follows:

$$x \text{ is grue} \overset{\text{definition}}{\equiv} \begin{array}{l} x \text{ is sampled and green} \\ \text{or unsampled and blue} \end{array}$$

Now, the argument

All sampled emeralds have been grue

$\therefore$ Probably, all emeralds are grue

conforms to the alleged rule of induction. However, that reasoning is definitionally equivalent to

All sampled emeralds have been green

$\therefore$ Probably, all sampled emeralds are green

and unsampled emeralds are blue

which we would surely reject. Instead, it would be our inductive practice to infer from the given information that unsampled emeralds are green. Thus, the schema is not accurate in general, although certain instances (for example, A = emerald, B = green) do produce acceptable reasoning. Therefore, we are left with the question: how to specify the class of predicates (so-called projectible predicates) whose substitution in the inductive schema will yield acceptable arguments.

(7) *Simplicity.* Given two incompatible theories which both accommodate our data, we feel that the simpler one is more likely to be true. The need for some such intuition arises in the first place because it is always possible to find various incompatible theories, all of which fit the evidence that we have already accumulated. A typical case of this phenomenon is the possibility of drawing many curves through our set of data points. Thus, suppose a scientist wants to know the functional relationship between two parameters X and Y (for example the temperature and pressure of a fixed quantity of some gas, confined to a chamber whose volume is constant). Let us say he can vary the value of X and measure the corresponding value of Y. Now, suppose that in this way he has obtained, for six values of X, the corresponding values of Y, and plotted these points on graph paper. The points turn out to lie upon a straight line; nevertheless, many other functions are compatible with them, as shown in Fig. 1.

Another example, familiar from the grue problem, involves conflicting hypotheses, each of which could accommodate the evidence:

All sampled emeralds are green

In this case the alternatives are as shown in Fig. 2. These equally account for our observations of green emeralds, though they diverge concerning the colour of future, as yet unexamined, emeralds.

This phenomenon – the prevalence of competing observationally adequate hypotheses – gives rise to three questions:

(A) How do we choose between the alternatives? On what is our preference based? Simplicity? If so, how is simplicity to be recognised? The

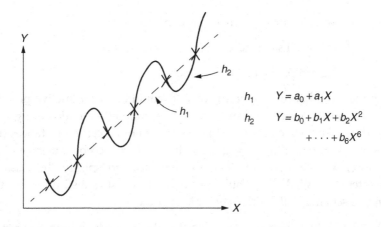

$h_1 \quad Y = a_0 + a_1 X$

$h_2 \quad Y = b_0 + b_1 X + b_2 X^2$
$$+ \cdots + b_6 X^6$$

Fig. 1

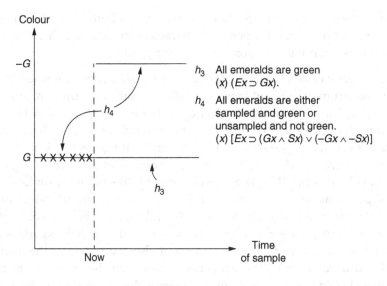

Fig. 2

grue problem is an element in this general problem of devising a description of our inductive practice.

(B) How is this measure of preferability to be combined with the knowledge that some theory fits certain data, to produce an overall assessment of its plausibility?

(C) What justifies our method of theory evaluation? Given an answer to question (A) – an account of the characteristics of a hypothesis which we take to recommend it above others which equally fit the data – what reason do we have to think that our procedure will tend to lead us towards the truth? In particular, why should we conclude, as we would given the evidence cited above, that probably all emeralds are green? This is the traditional problem of induction.

(8) *Ad hoc hypotheses.* The postulation of *ad hoc* hypotheses is thought to be somewhat disreputable. When an established theory is in danger of falsification by the discovery of facts it cannot explain, its proponents may patch up the theory in such a way as to reconcile it with the data. Such a manoeuvre is sometimes said to be *ad hoc* and scientists take a dim view of it. Consider, for example, the *ad hoc* claim that phlogiston has negative weight. This was proposed solely to rescue the theory that combustion of metals involves the escape of phlogiston, from the embarrassing observation that the ashes weigh more than the original metal. By reference to this and other examples we might hope to extract a definition of ad hocness, and a justification, based upon that characterisation, of the fact that we regard it as undesirable to postulate *ad hoc* hypotheses.

(9) *Diverse evidence*. We think that theories are better confirmed by a broad spectrum of different kinds of evidence than by a narrow repetitive set of data. Thus, intuitively, E_2 is better evidence for H than E_1 (see Fig. 3). Or, consider Snell's law of refraction: for any pair of media M_1 and M_2 there is a constant $\mu_{1,2}$ such that, for any i and r, if light is incident on the boundary between M_1 and M_2 at an angle i and is refracted at an angle r, then $\sin i / \sin r = \mu_{1,2}$ (Fig. 4). Some evidence for this claim is provided by 100 experiments in which r is measured for various different values of i, using the same pair of media M_1 and M_2, and found to be in accord with the claim. But stronger evidence is provided if the 100 experiments involve, not only a variation in i and r, but also a variation in the media used and a variation of the temperature at which the experiments are conducted. Again, what we want is a precise characterisation of breadth and an explanation of its evidential value.

(10) *Prediction versus accommodation*. There is an inclination to assume that a set of data provides better support for a theory if it was *predicted* by the theory before it was obtained, than if the theory was formulated after the data were obtained and was designed specifically to *accommodate* that information. This attitude is sometimes expressed in criticism of psycho-analytic theory. It is recognised that Freudians can concoct explanations afterwards of someone's behaviour, but it is felt that this is too easy – no

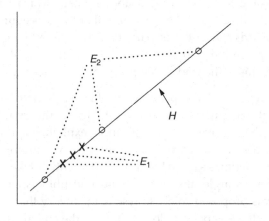

Fig. 3

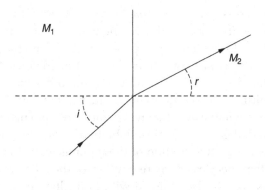

Fig. 4

real test of the theory. A real test would involve a definite prediction of what someone will do. Only then, and if the predictions are accurate, would we have good reason to believe the theory. We will explore the justification for this idea that accurate prediction has greater evidential value than the mere ability of a theory to accommodate, or subsume, or explain the data.

(11) *Desirability of further evidence.* When in doubt, it is prima facie desirable to acquire further evidence, for extra data tend to permit better assessments of the hypotheses under consideration. Imagine that two perfectly competent investigators are working independently of one another, but on the same problem – to decide between a set of alternative hypotheses in some area. Both scientists assess the plausibility of each hypothesis in the light of the data they have collected, and continually revise their assessments as new evidence accumulates. Now, suppose there is some time at which *A* has more information than *B*. *A* possesses all the facts known to *B*, but has some further results not known to *B*. And as a consequence of this difference they disagree about which is the most plausible hypothesis. In these circumstances we feel that *A* is in a superior position. His plausibility estimates, we suppose, are in some sense 'better' by virtue of the fact that they are based upon more information. Further- more, this intuition seems to be intimately related to our thirst for data. Faced with a sufficiently interesting problem, we are disposed to go to a great deal of trouble to gather as much relevant evidence as possible. The problem here is to explain these banalities. Why is extra information so desirable? Why is it that plausibility assessments based on more data are better? And what precisely is the meaning of 'better' in this context?

(12) *Realism versus instrumentalism.* One might naturally think that the point of science is to discover those fundamental, general and hidden, features of the world, which are responsible for observable phenomena. This is the so–called realist conception of scientific theories: that they are proposed as true descriptions of reality. On the other hand, it has been argued that realism fails to appreciate the inexorable pattern of scientific revolutions, the inevitable eventual renunciation of our current theoretical constructions and, therefore, the unsuitability of truth as the proper object of scientific theorising. Recognition of these points leads to instrumental-ism – the idea that theories are mere devices for the efficient organisation of data. But this view is also plagued with difficulties, of which the most serious is a need to distinguish sharply between the data statements, which correspond to facts in the world, and the theoretical formalism which is said to be designed merely to systematise them. I will suggest that the concept of subjective probability permits a sort of reconciliation between these views.

A taste of Bayesianism

As I said at the outset, one of our aims is a perspicuous representation of scientific methodology, and this project can be split into three intimately related components. First, we want clarification: precise formulations of the various intuitions and practices I have just described. Second, a sys-tematisation of these elements: we want to be able to formulate a set of fundamental principles and show that various, apparently independent, features of scientific methodology may be derived from them and reflect no more than an implicit commitment to those principles. Third, we would like to justify the methods and assumptions which underlie the way in which science is conducted. If we have succeeded in the first two tasks of clarifying and systematising our practice, then this third compon-ent will involve an attempt to justify those basic principles of which all other features of scientific methodology are consequences.

The orthodox, and most widely held, theory of scientific methodology is Hempel's hypothetico-deductive model. It is so-called because it supposes that the method of scientific investigation involves the stages shown in Fig. 5.

I think that this is approximately right, as far as it goes. But it is silent on a wide range of important matters. It fails to account for the testing and adoption of statistical hypotheses – for they don't entail observable pre-dictions. It provides no measure of the *degree* of confirmation conferred by

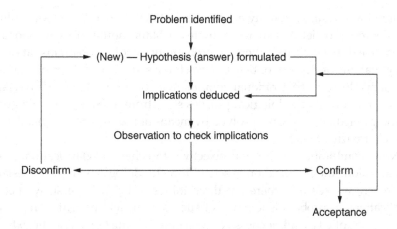

Fig. 5

a successful prediction: that is, no account of the relative evidential values of different pieces of evidence, and no account of which of two hypotheses should be preferred when they both predict the same data. And it gives no indication of when a hypothesis has been sufficiently confirmed to justify any confidence in it. Thus, the hypothetico-deductive model, as it stands, does not account for any of those items of scientific methodology we want to understand. We are going to need a theory which countenances degrees of confirmation.

So far, I haven't said anything particularly controversial. I've merely placed in a certain perspective some standard problems in the philosophy of science. But the approach I intend to take in trying to solve these problems *is* controversial. It is known as Bayesianism (after the eighteenth-century English mathematician, Thomas Bayes). And, broadly speaking, it derives from the view that the concept of subjective probability is of fundamental value in reaching an understanding of scientific methodology.

More specifically, the Bayesian approach rests upon the fundamental principle:

(B) That the degrees of belief of an ideally

rational person conform to the mathematical

principles of probability theory

For example, one principle of probability theory is

$$P(H) + P(-H) = 1$$

So it follows from (B) that my degree of belief that there is life on Mars plus my degree of belief that there is no life on Mars should, if I am rational, equal one. If (B) is correct, then the probability calculus constrains rational combinations of degrees of belief in much the same way that principles of deductive logic restrict rational combinations of full belief. The Bayesian idea is that our methodological puzzles stem from a fixation upon all-or-nothing belief, and may be resolved by means of the enriched probabilistic 'logic' of partial belief.

Now, combining our general objective (to achieve a clarification, systematisation, and justification of scientific methodology) with the Bayesian approach, we see that there are three things to be done: first, we need explications in probabilistic terms of such notions as 'confirmation', 'evidence', 'simplicity', 'ad hocness', etc.; second, we must examine the extent to which the various canons of scientific methodology may be derived from (B), and the extent to which further principles are required; third, we must evaluate the plausibility of (B). Can we provide some justification for it and for further assumptions which may be needed?

Let me end this chapter with an illustration of Bayesianism at work. It is an axiom of probability theory (Fig. 6) that

$$P(A\&B) = P(B)P(A/B)$$

$$\therefore P(H\&E) = P(H/E)P(E)$$

$$\text{and } P(E\&H) = P(E/H)P(H)$$

$$\text{but } P(H\&E) = P(E\&H)$$

$$\therefore P(H/E) = \frac{P(H)P(E/H)}{P(E)} \text{ (Bayes' theorem)}$$

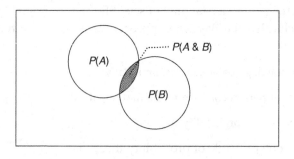

Fig. 6

Now, suppose we could establish (B). We could then infer that the degrees of belief $b(\ldots)$ of a rational person must satisfy the equation

$$b(H/E) = \frac{b(H)b(E/H)}{b(E)}$$

Now, suppose that H stands for some hypothesis and E for some evidence claim, and that H predicts E. Consider a rational person for whom the truth of E and of H are uncertain, that is,

$$0 < b(H) < 1$$
$$0 < b(E) < 1$$

We have

$$b(E/H) = 1$$

since it is an axiom of probability that if A entails B, then $P(B/A) = 1$. Thus, $b(E/H) > b(E)$. Therefore, employing Bayes' theorem, we find

$$b(H/E) > b(H)$$

In other words, rationality requires that the conditional degree of belief in H, given E, be greater than the degree of belief in H. Consider now a not wholly unnatural probabilistic explication of what it is for one statement to be evidence in favour of another, namely, the discovery of E would be evidence in favour of H if rationality requires that $b(H/E) > b(H)$. Given this clarification of the notion of evidence; it follows that the discovery of E would tend to confirm H; thus, we have supplied a rationale for the intuitive evidential value of accurate prediction.

Similarly, we can see why, in those circumstances, the discovery of $-E$ should disconfirm H. For it is a theorem of the probability calculus that

$$P(A/B) + P(-A/B) = 1$$

Therefore the beliefs of a rational person must satisfy

$$b(E/H) + b(-E/H) = 1$$

Therefore, in the given circumstances

$$b(-E/H) = 0$$
$$\therefore b(H/-E) = \frac{b(H)b(-E/H)}{b(-E)} = 0$$

Finally, we can account (pending a thorough treatment in Chapter 5) for the second item of scientific methodology discussed above: that relatively surprising predictions have greater evidential value. For it is natural to take $b(E)$ as an inverse measure of the surprisingness of E (that is, $S(E) = 1/b(E)$). Now suppose H entails E_1 and E_2. We have for a rational person

$$\frac{b(H/E_1)}{b(H/E_2)} = \frac{b(H)b(E_1/H)}{b(E_1)} \cdot \frac{b(E_2)}{b(H)b(E_2/H)}$$
$$= \frac{b(E_2)}{b(E_1)} = \frac{S(E_1)}{S(E_2)}$$

Thus, the more surprising the prediction E by H, the greater the value of $b(H/E)$ and the greater the difference between $b(H/E)$ and $b(H)$.

I will now turn to a general discussion of probability. The point of this sidetrack, as far as our project is concerned, is to clarify the notion of degree of belief, or what is generally called 'subjective probability' and to distinguish it from other probabilistic concepts. I will establish our fundamental thesis of Bayesianism (B): that rational degrees of belief conform to the probability axioms. And, in later chapters, I will employ this result in attempting to answer the various questions about scientific methodology outlined above.

2

Probability

The primitive theory

According to what I will call 'the primitive theory', the probability that a trial or experiment will have a certain outcome is equal to the proportion of possible results of the trial which generate that outcome. There are six possible results of throwing a die, and three of these yield an even number; so the probability of that outcome is 1/2.

A major fault with this definition is that it entails incorrect attributions of probability. The chances that a *biased* die will show an even number may not be 1/2. In order to pre-empt this objection, one natural strategy is to require that the possible results of the trial be equally likely. The probability, according to such a modified account, would be the proportion of equally likely results which generate the outcome. However, this saves the definition from incorrect consequences only on pain of circularity. The account is now inadequate as a definition of probability since it depends upon the notion of equally likely results. In order to apply the definition in the calculation of some probability, we must already grasp what it is for the alternative results of the trial to have the same probability.

A second deficiency of the modified primitive theory is that it applies only to a restricted class of those cases in which we attribute probabilities. For example, it does not encompass the claim that the probability of getting 1 or 2 with a certain biased die is 0.154. In many cases there may be no way to divide the possible results into a set of equally probable, mutually exclusive alternatives. For this reason, the primitive theory would also have difficulty in dealing with probability claims such as:

(a) The probability that Oswald shot Kennedy is 0.7.
(b) The probability, on the basis of evidence already collected by the NASA, that there is life on Mars is very slight.
(c) The probability that a radium atom will decay in any given ten-second interval is 0.001.

As we have seen, the primitive theory can be rescued from circularity only by some further characterisation of equiprobability. Note, however, that, even as it stands, the account is sufficiently substantial to determine all the mathematical axioms of elementary probability theory, namely:

(1) Probabilities are less than or equal to one
(2) If an outcome is necessary, its probability is one
(3) If outcomes q and s are mutually exclusive, then Prob (s or q) = Prob (s) + Prob(q)

There are four important ways of supplementing the primitive theory with an account of equal likelihood, and to each way corresponds a distinct concept of probability. In each case, equiprobability is explained in terms of a principle of indifference:

(1) The empirical principle of indifference: X and Y are equally likely results if there is no factor in the circumstances of the trial, which would tend to produce one result more frequently than the other.
(2) The subjective principle of indifference: X and Y are equally likely results, for a given person, if he has equal confidence in their occurrence.
(3) The rationalist principle of indifference: X and Y are equally likely results if there is no reason to have a greater expectation (degree of belief) that one rather than the other will occur.
(4) The logical principle of indifference: X and Y are equally likely results if the evidence confirms one to the same degree that it confirms the other.

Each of these principles is designed to remedy the first deficiency in the modified primitive theory – its circularity; but they all inherit the second deficiency – yielding concepts of probability which are undesirably restricted. They fail to illuminate those cases (such as the biased coin) in which there exists no set of equally probable, exhaustive and mutually exclusive, possible outcomes. However, for each principle there is a natural generalisation which accommodates this difficulty:

(1') The empirical interpretation: the probability that a certain kind of trial T will produce a certain sort of outcome O is the tendency, or propensity, of T to produce O, and is equal (in the classical version) to the limit of the relative frequency of O in an infinite sequence of T-trials.

From this viewpoint, a probability is an objective quantity whose values can only be determined by experimental investigation, for example by performing a long series of trials of type T. It gives a plausible account of the biased coin and radioactive decay. However, assertions like (a) The probability that Oswald shot Kennedy is 0.7, cannot comfortably be construed along these lines. Moreover, even if you are told that a coin is biased, you may nevertheless say that the probability of its coming up heads on the next toss is equal to 1/2. By this you do not mean that the relative frequency of heads would be 1/2, for you know that the coin is biased. Rather, your claim is related to the fact that you have absolutely no reason to expect heads rather than tails. Therefore, your claim should not be given an empirical interpretation. An alternative concept, applicable here, is given by a natural refinement of (2), namely:

> (2') The subjective interpretation: an attribution of some probability to a statement is an expression of the speaker's degree of confidence in its truth.

But are probabilities simply a matter of taste, as this suggests? A desirable element of objectivity is incorporated in

> (3') The rationalist interpretation: an attribution of some probability to a statement is a claim to the effect that, given the evidential situation, it is reasonable to believe, to the specified degree, that the statement is true.

Closely related to this interpretation, as we shall see, is one which is obtained in an analogous way from (4), namely,

> (4') The logical interpretation: statements of the form 'The probability of A relative to B is x' are claims about the degree to which the truth of B would support A.

Such claims are not empirically testable, but are allegedly logical truths expressing a sort of partial entailment relation between statements. They indicate (without explicitly expressing) the degree to which someone, whose total knowledge consists of B, should believe A.

I have given a brief sketch of some of the more influential attempts to characterise the meanings of probability judgements, and have tried to exhibit them as natural alternative extensions of a primitive account. In what follows I will go more carefully over some of the same ground, considering the legitimacy of those notions upon which each analysis depends (for example degree of belief, limit of relative frequency), and the range of its application. I begin with an examination of (2') and the theory known as subjectivism or personalism.

This section will develop the Bayesian framework for our subsequent treatment of scientific methods. It presents familiar arguments in favour of the existence of quantifiable degrees of belief, and in favour of the view that any rational combination of them must conform to the probability axioms. There will remain, however, the semantic question: what do we mean by attributions of probability? And to this I set forth standard considerations which suggest they should not be construed subjectivistically. Rather, it seems that such attributions are sometimes claims about what one *ought* to believe given the evidence (the rationalist interpretation); and sometimes claims about the objective value of a physical parameter – a propensity (the empirical interpretation). I will not be proposing anything very new or extreme in this chapter. In particular, I will not endorse any view to the effect that one particular concept of probability is fundamental. My aim is simply to prepare the ground for the methodological discussions which follow. A reader who is already versed in the philosophy of probability, and who is not disposed to scepticism about the Bayesian framework, might well skip this chapter and move on directly to the treatment of confirmation in Chapter 3.

Subjectivism

An attribution of some probability to a statement is, from this viewpoint, simply an expression of partial belief, to a certain degree, that the statement is true. When I say that the probability of heads is $1/2$, I mean solely that I have a degree of belief equal to one-half that the coin will land heads up. Degree of belief is measurable in terms of the betting odds which the subject would be prepared to offer on the truth of a proposition. Thus, if someone is offered the choice between 1.00, no strings attached, and 3.00 if and only if there is life on Mars, and is indifferent to which he receives, this would indicate a degree of belief of $1/3$ in life on Mars.

From this viewpoint, the mathematical laws of probability are not universally or necessarily true of degrees of belief. However, if a person is rational, he will distribute his probabilities – his degrees of belief – in accord with these laws. For only if he does this will he be able to avoid a so-called Dutch book being made against him. In other words, the betting odds which measure a person's probability judgements may, unless those judgements conform with the probability calculus, be taken advantage of by a suitable combination of stakes, in such a way that the person is bound to lose, whatever turns out to be the truth in those matters on which the bets are made. Suppose, for example, in violation of the law $P(A) + P(-A) = 1$, my degree of belief is $1/3$ that there is, and $1/3$ that there is not, life on Mars. That is to say, that I take 1.00 to be a fair stake in a bet whose prize is 3.00 if there *is* life on Mars; and that I am also indifferent between $1.00 and a prize of $3.00 if there is *no* life on Mars. Now, these commitments may be exploited by someone who bets me $1.00 that there is life on Mars, and another $1.00 that there isn't. He must win one of the bets. I have received a total of $2 in stakes, but I must pay out a $3 prize. Therefore, I am bound to lose $1.00, and so the odds I have offered, and the combinations of degrees of belief they reflect, are irrational. As we shall see, any such departure from the probability calculus is susceptible to this sort of criticism.

Thus, subjectivism is the sum of three distinct claims, and these should be evaluated separately:

(a) That people have degrees of belief.
(b) That rational degrees of belief should conform to the mathematical principles of probability theory.
(c) That probability assertions are expressions of degrees of belief.

Are there degrees of belief? There is no doubt that, roughly speaking, beliefs vary in strength. I am absolutely convinced that my pen is not black all over and white all over, not quite so certain that I was born in England, and even less sure, though I tend to believe it, that there are such things as neutrinos. I have no idea one way or the other about whether there is life on Mars. If so, I doubt that it is intelligent, and I would reject in the strongest possible terms the suggestion that I don't exist.

Degrees of belief are not only perfectly evident, but also conceptually valuable. Consider the following puzzle. Suppose there is to be a lottery among one million people for a single prize. You know that Smith's chances of winning are only one in a million and so you regard yourself as justified in believing that Smith will not win. This involves the principle

(a) S knows that $\text{Prob}(p) = \frac{1}{1000000}$
   $\rightarrow S$ is justified in believing $-p$

But, analogously, you know of every person $X_1, X_2, \ldots, X_{1000000}$ in the lottery that each of their chances of winning are equally $1/1\,000\,000$ and you can conclude with equal justification that none of them will win. But combining all of these million results produces the conclusion that no-one will win since

(b) (S is justified in believing p) $\wedge$
    (S is justified in believing q)
    $\rightarrow S$ is justified in believing ($p \wedge q$)

However, you know that this conclusion is false: there will certainly be some winner.

This problem – known as the lottery paradox – is neatly and quickly resolved if we introduce the idea of degrees of belief. For then we can deny (a) and replace it with the more plausible

(a') S knows that Prob $(p) = X$
   $\rightarrow S$ is justified in believing p to degree X

Now, in order to generate the paradox we would have to replace (b) with

(b') S is justified in believing p to degree X and
   S is justified in believing q to degree X
   $\rightarrow S$ is justified in believing $(p \wedge q)$ to degree X

But this is quite implausible. My degree of belief that neither X_1 nor X_2 will win the lottery should obviously be slightly less than my degree of belief that X_1 won't win.

Now this may give us a reason to hope that there is a coherent notion of degree of belief. But it does not establish that a precise quantitative concept can be made intelligible. We may well have an intuitive feeling for what would be meant by ascriptions of degrees of belief equal to 1 (certainty), 0 (certainty of denial), and even 1/2 (equal confidence). But what would be meant by saying that someone believes to degree 0.2 that there is life on Mars? How could the existence of such a degree be established?

It was suggested by a number of writers, Ramsey (1926), De Finetti (1937), von Neumann (1946), Keynes (1921) that we measure a person's degree of belief in a proposition by finding out the odds on which he would be prepared to bet on its truth. One natural way to implement this idea would be to define S's degree of belief in p, $b_s(p)$, as follows:

$$b_s(p) = x/y \text{ iff } [\$x] \overset{s}{\sim} [\$y \text{ iff } p]$$

Or, in English, the degree of belief of person S in the truth of p is x/y if and only if S would be indifferent given the choice between $\$x$ for certain and $\$y$ on condition that p is true. Consider, for example, the hypothesis, M, that there is life on Mars. S's degree of belief in it is given by:

$$b_s(M) = x/1 \text{ iff } [\$x] \overset{s}{\sim} [\$1 \text{ iff } M]$$

In other words, his degree of belief is that proportion x of $\$1.00$ which he would freely exchange for the privilege of $\$1.00$ to be paid if and only if there is life on Mars. As we would expect, if S is pretty sure that there is life on Mars, he will prefer the uncertain dollar unless x is relatively high.

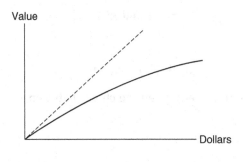

Fig. 7

On the other hand, to the extent that he strongly doubts the hypothesis, S will prefer the sure thing – $\$x$ – unless x is low. Thus, the proposed explication is quite faithful to our intuitive concept of degree of belief.

Unfortunately, there is a devastating objection to this proposal as it stands: namely, that the values people attach to different sums of money are not proportional to the amounts. Money has a diminishing marginal utility (value) (as is shown in Fig. 7) so that, for example, an extra $\$100$ is not worth as much to a millionaire as it is to the rest of us. This means that though S may be prepared to stake $\$5.00$ to win $\$10.00$ iff p, signifying $b_s(p) = 1/2$, he may not agree to stake $\$50.00$ to win $\$100.00$ iff p. And, in such a case, the proposed definition would dictate inconsistent attributions of degrees of belief to S.

If the definition is formulated in terms of units of value (as Ramsey intended) rather than in sums of money, it becomes immune to this objection. But now we are faced with the problem of defining these units – of specifying, for example, what it means to say that some amount of money, say $\$100.00$, would be twice as valuable (desirable) to S as $\$50.00$.

Ramsey's solution proceeds from the assumption that an agent maximises his expected gain. The so-called 'expected gain' of an action (decision, choice), is a technical notion designed to represent its desirability to the agent. The intuitive idea is that an action will have a high expected gain (utility, value) to the extent that there are possible consequences of it which seem to the agent to be both likely and desirable. Its magnitude is calculated by considering an exhaustive and mutually exclusive list of possible results of the action, multiplying the desirability of each possible result by its subjective probability, and adding these products together – thus obtaining a weighted averaged value of the possible consequences. In particular, the expected value, as far as S is concerned, of his decision to take

$$[\$c \text{ if } p \text{ and } \$d \text{ if } - p]$$

is

$$b_s(p)V_s(\$c) + b_s(-p)V_s(\$d)$$

So, if someone is indifferent given the choice between

$$[\$x]$$

and

$$[\$y \text{ iff } p]$$

then, from Ramsey's assumption that expected gain is maximised, we infer that those two options have the same expected value:

$$V[\$x] = V[\$y \text{ iff } p]$$

Applying the definition of expected value

$$V(\$x) = b(p)V(\$y)$$

This provides a rationale for the method of measuring degrees of belief mentioned above:

$$b_s(p) = V_s(\$x)/V_s(\$y) \text{ iff}[\$x] \overset{s}{\sim} [\$y \text{ iff } p]$$

Note how this would reduce to our original definition if the value of an amount of money were proportional to the amount.

In order to find out the subjective desirability of different sums of money, and thereby measure someone's degrees of belief, one may proceed as follows:

First, adopt the convention that the degree 1/2 will apply to any proposition which is believed to the same degree as its negation. In other words, that

$$\text{If } b(p) = b(-p), \text{ then } b(p) = 1/2$$

Second, find a proposition E such that the subject is indifferent between the options

$$[\$x \text{ if } E \text{ and } \$y \text{ if } - E]$$

and

$$[\$y \text{ if } E \text{ and } \$x \text{ if } - E]$$

although he is not indifferent between [$x] and [$y]. In that case, the expected utilities are equal, therefore

$$V(\$x)b(E) + V(\$y)b(-E) = V(\$y)b(E) + V(\$x)b(-E)$$

(leaving out the subscript S for the sake of convenience).

$$\therefore [V(\$x) - V(\$y)]b(E) = [V(\$x) - V(\$y)]b(-E)$$

But since the subject is not indifferent between [$x] and [$y]

$$V(\$x) - V(\$y) \neq 0$$
$$\therefore b(E) = b(-E)$$
$$\therefore b(E) = 1/2$$

Once we have found a proposition E, such that S is indifferent between those two options, and which is, therefore, believed to degree $1/2$, then we are in a position to determine if the subjective value of some amount $\$y$, is, or is not, half-way between the values of two other amounts, $\$x$ and $\$z$. For if S is indifferent between

$$[\$y]$$

and

$$[\$x \text{ if } E \text{ and } \$z \text{ if } - E]$$

then their expected values are equal; therefore

$$V(\$y) = V(\$x)b(E) + V(\$z)b(-E)$$
$$= \frac{1}{2}[V(\$x) + V(\$z)]$$

Third, stipulate that $V(\$0)$ be 0 vals and $V(\$100)$ be 100 vals (in much the same way that the Centigrade scale is defined, by conventionally specifying how many degrees are to attach to the freezing and boiling points of water).

Fourth, determine which amount is worth 50 vals to S, then which amounts are worth 25 vals, 75 vals, 150 vals, and so on, by means of the method just described. Simply find that sum of money $\$y$, such that he is indifferent between

$$[\$y]$$

and

$$[\$100 \text{ if } E \text{ and } \$0 \text{ if } - E]$$

Since their expected values are equal

$$V(\$y) = V(\$100)b(E) + V(\$0)b(-E)$$
$$= 100 \times \frac{1}{2} + 0 \times \frac{1}{2}$$
$$= 50 \text{ vals}$$

Similarly (supposing we have discovered y), that sum of money $\$w$ such that S is indifferent between

$$[\$100]$$

and

$$[\$w \text{ if } E \text{ and } \$y \text{ if } - E]$$

is worth 150 vals to S.

Fifth, once we have attached values to items by some such procedure, we can employ the principle

$$b(p) = V_1/V_2 \text{ iff } [V_1] \sim [V_2 \text{ iff } p]$$

to measure degrees of belief.

Of course, this method of measurement is not infallible. In describing and justifying the procedure, I have glossed over several idealising assumptions, for example that the subject is not constitutionally averse to risk. If he *does* put a negative value on risk itself then a reaction of indifference to the choice between, say,

$$[5 \text{ vals}] \text{ and } [10 \text{ vals iff there is life on Mars}]$$

would not reflect a degree of belief of $1/2$ in the statement. For the expected utility of the second option is

$$10b(M) - \varepsilon$$

where ε is the undesirability of the risk. Thus, we may conclude merely that

$$5 = 10b(M) - \varepsilon$$
$$b(M) = \frac{5 + \varepsilon}{10}$$

which is greater than would be suggested if the risk aversion were not taken into account. However, the fallibility of our method of measurement does not reflect badly upon the notion of degree of belief. As with other concepts, the search for an operational definition is both hopeless and unnecessary. Sufficient clarification of the concept is provided by its role in the principle of expected utility. The practical measurement problem, though it might well be handled along the lines indicated above, can be unsurprisingly messy when various distorting factors come into play.

This completes my defence of the first component of subjectivism: there are indeed quantifiable, measurable degrees of belief. Let us now consider the second claim – first proved by De Finetti – that combinations of such beliefs should, if reasonable, conform to the probability calculus. The axioms of elementary probability theory are:

(1) Prob $(p) \leq 1$
(2) Prob (tautology) = 1
(3) If p and q are mutually exclusive, then $\text{Prob}(p \vee q) = \text{Prob}(p) + \text{Prob}(q)$

Thus, in order to establish the second component (b) of subjectivism (what I referred to in the first chapter as the fundamental thesis of Bayesianism), it is necessary to show that rational degrees of belief, as characterised above, must conform to the principles:

(1') $b(p) \leq 1$
(2') b (tautology) = 1
(3') If p and q are mutually exclusive, then $b(p \vee q) = b(p) + b(q)$

First, if $b(p)$ were $1 + E$ (E positive), then $[1 + E$ vals$] \sim [1$ val iff $p]$ vals; but it would be irrational to stake $(1 + E)$ vals in a bet which could return 1 val at most.

Second, I shall simply assume that an ideally rational person can and does establish the truth of all logical truths. Clearly we are by no means ideally rational in this sense. However, in the context of science and ordinary life, with which we are concerned, the relevant logical truths are sufficiently elementary that this idealisation does not diverge significantly from our actual capacities.

Third, suppose that $A_1, \ldots , A_k$ are a set of alternatives of which it is known that one, and only one, is true. Then a rational person will be indifferent between

$$[x \text{ vals}]$$

and the combination of gambles

$$\{[x \text{ vals iff } A_1] \text{ and } [x \text{ vals iff } A_2], ...,$$
$$\text{and } [x \text{ vals iff } A_k]\}$$

Therefore, if S is rational, and if he knows that p and q are jointly impossible then he will know that p, q and $-(p \vee q)$ are a set of exhaustive, mutually exclusive outcomes. So

$$[1 \text{ val}] \sim \{[1 \text{ val iff } p] \text{ and } [1 \text{ val iff } q]$$
$$\text{and } [1 \text{ val iff } - (p \vee q)]\}$$

Also, S knows that $(p \vee q)$ and $- (p \vee q)$ are a set of exhaustive, mutually exclusive outcomes. So

$$[1 \text{ val}] \sim \{[1 \text{ val iff } p \vee q] \text{ and } [1 \text{ val iff } - (p \vee q)]\}$$

Therefore, combining these results,

$$\{[1 \text{ val iff } p \vee q] \text{ and } [1 \text{ val iff } - (p \vee q)]\}$$
$$\sim \{[1 \text{ val iff } p] \text{ and } [1 \text{ val iff } q] \text{ and } [1 \text{ val iff } - (p \vee q)]\}$$

Now, removing the same benefit, namely [1 val iff $-(p \vee q)$], from both sides, we obtain

$$[1 \text{ val iff } p \vee q] \sim \{[1 \text{ val iff } p] \text{ and } [1 \text{ val iff } q]\}$$

But from our initial characterisation of degree of belief we have

$$b(G) = x \quad \text{iff}[x \text{ vals}] \sim [1 \text{ val iff } G]$$
$$\therefore \ [b(G) \text{ vals}] \sim [1 \text{ val iff } G]$$

Thus, substituting $(p \vee q)$, p, and q respectively for G, we obtain

$$[b(p \vee q) \text{ vals}] \sim \{[b(p) \text{ vals}] \text{ and } [b(q) \text{ vals}]\}$$
$$\therefore b(p \vee q) = b(p) + b(q)$$

Alternatively, a Dutch book argument may be given for the case of (3′). Suppose p and q are mutually exclusive. If $b(p \vee q) = b(p) + b(q) + E$, then $[(b(p) + b(q) + E)\text{vals}] \stackrel{S}{\sim} [1 \text{ val if } (p \vee q)]$. Suppose E is positive. Then a shrewd bettor could stake $b(p)$ vals with S on the truth of p (standing to gain 1 val if p is true), and also stake $b(q)$ vals (standing to gain 1 val) on the truth of q, and he could simultaneously get S to stake $[b(p) + b(q) + E]$ vals (standing to gain 1 val) on the truth of $(p \vee q)$. Now, whatever happens, S must lose E vals. For if either one of p or q is true, S's total outflow is $[1 + b(p) + b(q) + E]$, and his income is $[1+b(p) + b(q)]$. And if neither is true, his outflow is $[b(p) + b(q) + E]$ and his income merely $[b(p) + b(q)]$. (It is impossible for both p and q to be true.) Similarly, if E is negative, the bettor can again exploit the odds by getting S to stake $b(p)$ and $b(q)$ vals on the truth of p and q respectively, whilst simultaneously staking $[b(p) + b(q) + E]$ vals with S on the truth of $(p \vee q)$.

We have seen that there is considerable merit in two of the three elements of subjectivism: people have degrees of belief and the degrees of belief of a rational person must conform to the probability calculus.

It remains to examine the third element: that a probability assertion is merely an expression of the speaker's degree of belief.

This view encounters a number of difficult questions: How may it be reconciled with the existence of disagreement? If divergent probability claims each report the speakers' state of mind, there would seem to be no conflict between the assertions. How does this view accommodate the presence of critical standards, and the large measure of intersubjective agreement which derives from the general recognition of those standards? How can it explain the scientific and practical value of probabilities – their ability to yield fruitful predictions and explanations and facilitate the profitable operation of insurance companies?

Such problems threaten the idea that probability claims should ever be given a purely subjectivist interpretation. However, there is, in addition, a conclusive reason for denying that all probability claims are subjective. We do have the idea of the unknown probability that a coin will land heads. This probability is something to be discovered by doing an experiment, and it is irrelevant to its actual value what my degree of belief happens to be that it will land heads. It seems clear that empirical probability claims of this type should not be given a subjectivist interpretation.

As for other claims, the existence of disagreement and critical standards suggest that a rationalist or logical interpretation is appropriate. When someone states that the probability is $1/4$ of there being life on Mars, this is assessed in the light of the evidence as being a rational or irrational assertion. Underlying any criticism or approval of such a claim, must be views about whether or not it is objectively the case that the available evidence justifies that degree of belief in the proposition that there is life on Mars. Now a subjectivist may accept this, conceding that the probability claim is somehow based upon some underlying view of the objective degree to which the evidence makes it rational to believe the proposition. Yet he may still maintain that the claim merely expresses the speaker's degree of belief.

But we are engaged in a lost cause. No-one denies that 'It is raining' normally indicates a belief on the part of the speaker. He would not need to add 'I believe that it is raining'. However, no-one would contend that the latter statement is part of the truth conditions of the former – what is explicitly asserted. Similarly, the claim 'The probability of p is x' may be taken to indicate a certain degree of belief on the part of the speaker, though this belief is not entailed by the truth conditions of the probability claim. Once a subjectivist concedes the existence of an objective basis for probability claims, it is natural, and squares with our feelings about other kinds of assertion, to take the probability statement as an assertion about the presence of objective conditions, and to regard the degree of belief which is indicated as a mere implicature of the claim. I therefore reject the subjectivist analysis of the meaning of probability assertions. But I accept the rest of subjectivism – degrees of belief and the probabilistic condition of rationality. These will provide the basis for our account of scientific methodology.

Before turning to the rationalist interpretation of probability – a view which inherits the virtues but avoids the flaws of subjectivism – I want to introduce a derivative subjectivist notion which will play a substantial role in the following chapters. This is the idea of a *conditional* degree of belief, written $b(p/q)$, and defined as follows:

$$b(p/q) = \frac{b(p \wedge q)}{b(q)}$$

As its name suggests, a conditional degree of belief is taken to represent the degree of belief in p relative to the supposition that q is true. To see that this is justified – that $b(p/q)$, as defined above, may be appropriately regarded in this way as a conditional degree of belief – we must show that the magnitude of $b(p/q)$ will reflect a willingness to make a conditional bet on p, in just the way that the magnitude of $b(p)$ reflects a willingness to make an unconditional bet on p. Thus, S would stake $b_s(p)$ vals for unit gain iff p is true. Similarly, as I will now argue, $b(p/q)$ vals would be staked for unit gain iff p is true – subject to the further condition that all bets are off unless q is true. To see this, note that:

$$b(q) = 1/x \text{ iff } [1 \text{ val}] \sim [x \text{ vals iff } q]$$

that is,

$$[1 \text{ val}] \sim [1/b(q) \text{ vals iff } q]$$

Similarly

$$[1 \text{ val}] \sim [1/b(p \wedge q) \text{ vals iff } p \wedge q]$$

Therefore, by transitivity of indifference,

$$[1/b(q) \text{ vals iff } q] \sim [1/b(p \wedge q) \text{ vals iff } p \wedge q]$$

$$\therefore \left[\frac{b(p \wedge q)}{b(q)} \text{ vals iff } q \right] \sim [1 \text{ val iff } p \wedge q]$$

And now, given the definition of b(p/q)

$$[b(p/q) \text{ vals iff } q] \sim [1 \text{ val iff } p \wedge q]$$

Thus, $b_s(p/q)$ is the ratio of stake to gain which S would take to be fair, in a bet on the truth of p, which is agreed to be operative only on condition that q is true. So it *is* appropriate to regard $b(p/q)$ as a conditional degree of belief. For example, suppose

$$[3 \text{ vals}] \overset{s}{\sim} [10 \text{ vals iff there is life on Mars}]$$

and

$$[5 \text{ vals iff there is life on Venus}]$$
$$\overset{s}{\sim} [10 \text{ vals iff there is life on Mars and Venus}]$$

S is prepared to stake more on the existence of life on Mars if the bet is to be conditional upon there being life on Venus: his degree of belief in life on Mars is increased by the supposition of life on Venus.

In having a conditional degree of belief $b(p/q) = x$, S constrains what he may reasonably predict would be his degree of belief in p if he were to discover the truth of q. More specifically, insofar as S will make any commitment at all to what his degree of belief in p would be, then reason dictates that he expects it to be x. Otherwise, if it were, say, $x + \varepsilon$, then this belief could be exploited in the following manner: by virtue of his conditional degree of belief,

$$[x \text{ vals iff } q] \overset{s}{\sim} [1 \text{ val iff } p \text{ and } q]$$

and, by virtue of his hypothetical expectation, he guarantees that

$$\text{if } q \text{ is discovered, then } [\{x + \varepsilon\}\text{vals}] \overset{s}{\sim} [1 \text{ val iff } p].$$

Now we give S the option $[x \text{ vals iff } q]$ in return for $[1 \text{ val iff } p \wedge q]$. If q is not discovered, then nothing happens. But if q is discovered, we then hold S to his guarantee and give him $[1 \text{ val iff } p]$ in return for $[(x + \varepsilon) \text{ vals}]$. Therefore, S stands to gain nothing from these exchanges; but if q is discovered he loses ε vals. This argument depends upon the assumption that ε is positive. If it is negative, simply make the opposite exchanges, and again it follows that S cannot win, but might well lose ε vals. Thus, a conditional degree of belief $b(p/q) = x$ ought not to be combined with a simultaneous commitment to believe p to degree y if q is discovered, *unless* $y = x$.

Notice, however, that $b_s(p/q) = x$ does not entail anything either about what S's degree of belief actually will be if q is discovered, or about what it ought to be. For S may adopt the conditional degree of belief without committing himself in advance to having, hypothetically, a certain degree of belief in p should q be discovered. But only if he does adopt a commitment can we show that reason constrains his eventual belief. Furthermore, even given the existence of such a prior commitment, a change of mind is perfectly possible, and so the hypothetical belief claim can easily turn out to be incorrect.

Thus, I do not subscribe to the so-called conditionalisation principle, advocated by many Bayesians, whereby $b_s(p/q) = x$ implies that the discovery of q should yield $b_s(p) = x$. In the first place, as shown above, there is no Dutch book rationale for conformity to it. Second, it is difficult to find a plausible version of the principle: clearly there are circumstances where q might be discovered in which a change of belief in p to x would not be appropriate. Third, we are trying to obtain conditions for the rationality of systems of beliefs. The conditionalisation principle suggests, wrongly in my opinion, that the reasonability of a theoretical hypothesis should depend upon something more than whatever concurrent further beliefs may be adduced in support of it. See pp. 74–7 for a general discussion of this point.

The rationalist interpretation

One natural response to the defective part of subjectivism – its implausible construal of probability claims – is the adoption of a rationalist interpretation. In this view, an attribution of probability to a statement does not describe the speaker's personal degree of belief in the statement, but makes an objective claim to the effect that it is *reasonable* to believe, to the specified degree, that the statement is true. This accounts for the existence of genuine disagreement over probabilities, and the criticism and revision of probability assertions in the light of recognised standards. However, it remains to be seen what does make a given degree of belief rational or irrational – to formulate principles which will allow us to determine which degrees of belief are rational.

It is useful to distinguish two forms of rationalism. We may call them weak and strong rationalism. In the first view, the only condition which must be met for a combination of degrees of belief to be rational is that they be coherent, that is, conform to the probability calculus. Thus, according to a weak rationalist, 'The probability of p is x' means 'Given the available evidence, anyone's degree of belief will conform to the probability calculus only if their degree of belief in p is x'.

A strong rationalist holds, on the other hand, that rationality constrains the legitimate combinations of degrees of belief which a person may have, but not only by the requirement that they conform to the probability calculus. According to strong rationalism, rationality not only requires coherence, but is also governed by further principles which regulate partial belief in hypotheses, given the evidential situation. Thus, given an evidential situation in which all that is known is

(a) The proportion of Fs which are G is x
(b) k is an F

it might be claimed that rationality requires a degree of belief equal to x in the statement that k is G. This view may be expressed as the assertion

Prob (k is G/k is F and the proportion of Fs which are G is x) = x

Thus the research programme of a strong rationalist will include an attempt to formulate principles such as this; or, more ambitiously, to provide a way of determining, for any statements, p and q, the degree to which it is rational to believe p in an evidential situation in which all that is known is the truth of q.

At first sight it seems that the strong version of rationalism is more plausible than the weak version. For it seems that coherence, though a necessary condition for rationality, would hardly be sufficient.

For example, suppose I am confronted with an urn which I know contains 1 red ball and 99 white balls. I am planning to reach in, rummage around without looking, and select one of the balls. What degree of belief ought I to have that I will pick the red ball? The correct answer would seem to be about 1/100. Yet mere coherence would not appear to determine that result. I could insist that the probability of getting the red ball is 3/4 and still maintain coherence provided my other degrees of belief are adjusted accordingly. So my degree of belief that I will pick a white ball would have to be 1/4.

Or, suppose all I know is that there are 100 balls, some white and some red, and that I have picked out 50 at random and found 40 red and 10 white. In the light of that evidence it seems clear that I ought to have a degree of belief greater than 1/2 that most of the remaining balls are red. But again, mere coherence would not seem to dictate this conclusion.

However, these considerations are not conclusive. The weak rationalist has an ingenious reply up his sleeve – though it is not, in the final analysis, quite adequate. His reply rests upon a phenomenon called 'the swamping of the priors' – an overwhelming of preconceptions by new data, and it is intended to show that the constraint of coherence is more powerful than one might first suppose.

Consider two hypotheses, H_1 and H_2. We have Bayes' theorem:

$$P(H/E) = \frac{P(H)P(E/H)}{P(E)}$$

$$\therefore \frac{P(H_1/E)}{P(H_2/E)} = \frac{P(H_1)}{P(H_2)} \cdot \frac{P(E/H_1)}{P(E/H_2)}$$

Now, suppose two people diverge in the degrees of belief they assign to H_1 and H_2. They may nevertheless be in fairly close agreement about their relative plausibility, given evidence E, for it may be

(a) that they agree about $P(E/H_1)$ and $P(E/H_2)$ and
(b) that $P(E/H_1)/P(E/H_2)$ is either very large or very small compared with either of their estimates of $P(H_1)/P(H_2)$

Consequently, their prior views about the relative plausibility of H_1 and H_2 are swamped by the extreme value of $P(E/H_1)/P(E/H_2)$, they may come to agree about which hypothesis is most probable in the light of E even though they disagreed initially.

For example, suppose we pick a coin which we know to be either fair, or biased $3/4$ in favour of heads. Let H_1 be 'The coin is fair' and H_2 be 'The coin is biased $3/4$ towards heads'. Now, suppose we toss the coin 100 times and get 75 heads. This is E and E is equivalent to a disjunction of mutually exclusive possible sequences – where each possibility specifies which particular tosses yield heads. Therefore, the probability of E is equal to the sum of the probabilities of these possible sequences – the sum of the probabilities of each possible sequence which arranges the 75 heads and 25 tails in some particular way. Assuming the result of each toss is probabilistically independent of the others, it follows that the probability of any such particular sequence is $x^{75}(1 - x)^{25}$, where x is the probability of heads in a single toss.

$$\therefore P(E/H_1) = (1/2)^{75}(1/2)^{25} \times \text{number of possible sequences}$$

$$\text{and } P(E/H_2) = (3/4)^{75}(1/4)^{25} \times \text{number of possible sequences}$$

$$\therefore \frac{P(H_1/E)}{P(H_2/E)} = \frac{P(H_1)}{P(H_2)} \cdot \frac{(1/2)^{75}(1/2)^{25}}{(3/4)^{75}(1/4)^{25}}$$

$$\approx \frac{P(H_1)}{P(H_2)} \cdot \frac{1}{2^{20}}$$

Thus, even those who disagree radically about the prior probabilities (priors) of H_1 and H_2 should come to agree, given evidence E, that H_2 is more probable.

Despite this result, there are reasons to suspect that the swamping of the priors will not be sufficient to dispel the impression that there is more to rationality than mere coherence. Firstly, we can see in the above example that the priors are swamped only if $P(H_1)/P(H_2)$ is not greater than 2^{20}. Therefore, it is rational to take the evidence E as justifying $b(H_2) > 1/2$ only if it would have been irrational to suppose, in advance of the data, $b(H_2) = 2^{-20}$. Yet it is hard to see how a weak rationalist could show this. Furthermore, agreement is assumed that the conditional degrees of belief in heads given, H_1 and H_2, are $1/2$ and $3/4$ respectively. Yet no account has been given of how these commitments may come to be required. Also, consider the set of cases in which all someone knows is that the proportion of As which are B is x, and that k is an A. How can the weak rational subjectivist show that the reasonable degree of belief in the statement 'k is B' would be x? Finally, there are cases in which the evidence will not swamp the priors – namely, when the available data E are *entailed* by both of the competing hypotheses. Thus, let H_1 be 'All emeralds are green', H_2 be 'All emeralds are green before the year 2000 and blue afterwards', and E be 'If 1000 emeralds will have been examined prior to the year 2000, then all of them will have been found to be green'. Now we have

$$\frac{P(H_1/E)}{P(H_2/E)} = \frac{P(H_1)}{P(H_2)} \cdot \frac{P(E/H_1)}{P(E/H_2)}$$
$$= P(H_1)/P(H_2)$$

Thus, the posterior relative probability of H_1 and H_2 is just the same as their prior relative probability. The evidence E does nothing to help us choose between them. Moreover, it is easy to extrapolate from this result and see that, in general, the swamping of the priors phenomenon will not account for our intuitions of rationality – our judgements about which theory should be preferred – in that pervasive class of cases which involve the choice between incompatible, empirically adequate hypotheses. Certainly, the onus of proof is with the weak rationalist to explain how, from mere coherence, these various intuitions may be extracted.

Whichever version of rationalism is adopted, the meaning of a probability claim is taken to be

One ought to believe p to degree x

Now, it is plain that the degrees of belief it would be rational to have in a statement such as 'There is life on Mars', depends upon the available evidence. 'The probability of life on Mars is low' may be true relative to certain evidential conditions, and false relative to others. Consequently, underlying such categorical probability claims, there are conditional probability claims of the form:

> In evidential conditions E, the probability of p is x

or, in other words,

> In evidential conditions E, one ought to believe p to degree x.

The logical interpretation

It is now a short step to Carnap's (1950) logical interpretation of probability. Assume that one's evidential circumstances may be represented by knowledge of a single proposition E – the conjunction of observed facts. Then a rationalist conditional probability claim has the form

$$\text{Prob}(p/E) = x$$

Now compare this with a statement of the form

$$E \rightarrow p$$

That implies a constraint upon any rational system of beliefs – namely, that it would be irrational to believe that E is true and p is false. However, we do not take it that '$E \rightarrow p$' *means* 'It would be irrational to believe that E is true and p is false'. Rather, we construe '$E \rightarrow p$' as expressing a logical relation of entailment between E and p, asserting the impossibility of E being true without p being true as well. Similarly, we should recognise the existence of probability statements of the form, E confirms p to degree x, $c(p/E) = x$, expressing an objective relation of probabilification or partial entailment between p and E, implying, but not directly asserting, that someone whose total evidence is E should believe p to degree x. To each rationalist probability claim about belief:

> Given total evidence E, one should believe p to degree x

There will correspond a logical probability relation between statements:

> E confirms p to degree x

Carnap's programme for inductive logic was to devise a function

$$c(p/E) = x$$

which would specify, for any statements p and E, the degree to which E confirms p or, in other words, the logical probability of p given E. This function would thereby determine the degree to which p should be believed, given total evidence E.

Carnap offers a number of reasons for calling this conception 'the logical interpretation of probability':

First, the relation of 'probabilification to degree x' holds between statements and may be regarded as a weak version of the logical relation of entailment.

Second, such probability statements are supposedly knowable *a priori*, and their truth values determined solely by the meanings of constituent terms. If it is a fact that

Given total evidence E, one ought to believe p to degree x

then we would be obliged to recognise this fact if we were in the state of total evidence E. But in that case we should be able to see, even now, that in those hypothetical circumstances we would recognise it. So we should see now, regardless of our present evidence, that the right degree of belief in p, given total evidence E, is x. Consequently, the statements of rational and logical probability are not empirical.

Third, the function c is *defined* by specifying, for any pair of statements, the value of the function. Thus, the truth value of $c(p/E) = x$ is just a matter of definition – it may be discovered simply by consulting the definition of c and establishing whether or not the triple $\langle p, E, x \rangle$ conforms to it.

Fourth, whether or not E logically entails p depends solely upon the logical forms of the statements E and p. Similarly, Carnap thought, the degree to which E logically probabilifies p should depend on just these formal characteristics. Thus, he hoped to devise a confirmation function c with the following property: that if the two pairs of statements $\langle E_1, p_1 \rangle$ and $\langle E_2, p_2 \rangle$ have the same logical form, then

$$c(p_1/E_1) = c(p_2/E_2)$$

For example, the numerical value of

c (All As are B/Objects $k_1, k_2, \ldots, k_n$ are both A and B)

will not depend upon which particular predicates are substituted for A and B.

I have simplified the preceding discussion by suggesting that there is a unique, correct c-function which specifies sharp numerical probabilities for each statement, given the total evidence. In fact, Carnap recognises that there may, given the evidence, be a range of degrees of belief in, say, life on

Mars, which would not be irrational. Perhaps reason dictates only that the probability is low, and does not require a specific degree of belief. If this is so, then the logical probability of p, given E, will be a set of numbers I, such that, given total evidence E, one's degree of belief in p may be some number inside I and may not be outside I. And in that case the primary object of research in inductive logic becomes a characterisation of all the permissible c-functionals – all the number-valued confirmation functions which yield permissible degrees of belief.

The evidential state

Both rationalism and the logical interpretation involve some distinction between the evidential state of a subject and the further beliefs to be justified on the basis of the information which constitutes that state, and it is incumbent upon a proponent of one of these views to say something about how this distinction is to be drawn – how the subject's evidential state is to be specified.

A natural strategy (adopted, for example, by Carnap) is to invoke the difference between observation sentences and theoretical sentences. The subject's evidential state is said to consist in those beliefs acquired non-inferentially, directly on the basis of observation. Carnap supposes, moreover, that there is a class of permissible confirmation functions, $c(H/E) = x$, from pairs of statements to real numbers, each of which specifies a degree of belief x, which it would be rational for a person to have in H, whose evidential state consists in the conjunction E of observational beliefs. Thus, to determine whether or not a subject's total system of beliefs is rational, we first identify those elements which comprise his evidential state; second, we conjoin beliefs into a single statement E; and finally, we look through the class of permissible c-functions to see if there is one which prescribes, relative to E, exactly those degrees of belief which are held by the subject. If, and only if, there is one, the system of beliefs is rational.

I want to focus on three difficulties with this idea. In the first place, it presupposes that observation beliefs are held with certainty – to degree 1. Richard Jeffrey (1965) has emphasised the implausibility of this assumption. He points out that there are poor circumstances for the acquisition of data, bad light, for example, in which the immediate results of observation may well be unsure and tentative beliefs. We should accommodate the possibility that our perceptual beliefs are not all inferred from absolutely certain sense-data judgements. Therefore, the evidential circumstance cannot always be represented by a statement. Instead, it must be considered in general to be a set of subjective probability assignments, such as

$$p_1, q_{1/2}, r_{1/3}, s_1, t_{1/4}$$

where each statement is believed to the degree indicated by its subscript.

Now the question arises as to whether a Carnapian c-function will suffice to determine probabilities, given such evidential states. Jeffrey does provide the following plausible and helpful principle:

$$c(p/E_x) = xc(p/E) + (1 - x)c(p/-E)$$

But this doesn't solve the whole problem. For how are we to treat evidential circumstances, such as the one schematised above, which include several more or less uncertain observations? Suppose someone has tentative beliefs in E and F to degree x and y respectively. Jeffrey's rule would suggest that the credibility of H be given by

$$P(EF)c(H/EF) + P(E\overline{F})c(H/E\overline{F})$$
$$+ P(\overline{E}F)c(H/\overline{E}F) + P(\overline{E}\,\overline{F})c(H/\overline{E}\,\overline{F})$$

Now, if E and F are thought to be probabilistically independent, then this formula may be employed. For, in that case, $P(EF) = xy$, $P(E\overline{F}) = x(1 - y)$, $P(\overline{E}F) = (1 - x)y$ and $P(\overline{E}\,\overline{F}) = (1 - x)(1 - y)$. But if not, we have no way to compute $P(EF)$, $P(E\overline{F})$ etc. In these circumstances the rule is applicable, only if we suppose that the evidential state itself contains not just E_x and F_y but also degrees of belief in all the combinations EF, $(E\overline{F})$ and so on: it must take the form

$$E_x, F_y, (EF)_z, (E\overline{F})_w \ldots$$

Thus, Carnap's c-functions will determine probabilities or, what ought to be believed, even given evidential circumstances which involve unsure observation: but only provided the evidential state consists in the possession of specified degrees of belief in a set of exhaustive and mutually exclusive statements. We may then define a broader class of permissible d-functions, $d(H/a) = x$, each of which specifies a degree of belief x which it would be rational for a person to have in H, whose evidential state a consists in a certain combination of degrees of belief of the required kind. This precludes a general logical construal of English probability attributions, for the probability relation between a and H is not a relation between statements. However, it is perfectly compatible with the existence of *some* logical probability statements; since in some cases a does consist in the certain belief of a single statement.

The second problem is deeper. It concerns the very legitimacy of Carnap's distinction between evidential state and further beliefs or, in his view, between data and theory. On the Carnapian approach under consideration, the rationality of those certain, observational beliefs which constitute a subject's evidential state, cannot be called into question. What can be scrutinised is merely the rationality, relative to this state, of further (theoretical) beliefs. Thus, this picture is a form of foundationalism: a system of beliefs divides into, first, a basis of intrinsically credible observations which do not stand in need of justification and, second, a set of conclusions which may or may not be appropriate and which therefore introduce the possibility of irrationality. However, as many philosophers have noted, observation may be theory laden: no belief is immune from revision and none can stand permanently above the need for justification. It can happen that beliefs based upon observation are reasonable only by virtue of the fact that the subject embraces certain theories, concerning, for example, the reliability of his perceptual apparatus in the circumstances in which the observation was made, and the nature of the phenomenon under observation. Thus, one speaks of seeing bacteria under a microscope, and this claim presupposes theoretical beliefs in the fields of optics and biology. Despite the optimistic and continual efforts of positivistically inclined philosophers, it has not proved possible to devise or isolate a theory-neutral language in which the results of pure observation may be recorded and used in the arbitration of all theoretical disputes. If this enterprise is, as it seems, misguided and hopeless, it would appear that a more holistic picture is called for, in which the notion of intrinsically credible observation reports is abandoned.

Another substantial objection to Carnap's approach, emphasised by Putnam (1962) and Kuhn (1962), is the fact that the probability of a hypothesis seems to depend on more than just the data, the hypothesis in question, and our background theories. Often the probability of a hypothesis – the degree of belief it should be given – will change simply because someone is sufficiently ingenious to think of a good alternative. The credibility of classical mechanics was diminished by the mere elaboration of relativity theory. Thus, it matters to the credibility of a hypothesis whether alternative theories have been formulated and proposed. The relevance of this factor can perhaps be accommodated by Carnapian c-functions, by allowing beliefs in the *existence* of theories to constitute elements of the evidential state.

What now becomes of the rationalist and logical interpretations of probability judgements? They survive these difficulties, but only by tolerating a vaguer and contextually dependent notion of 'evidential state'. A rationalist view could be developed along the following lines: in any situation in which a probability claim is asserted, some set of beliefs is taken (often tacitly) to constitute the evidential state. This set, which can include partial beliefs, may sometimes be explicitly specified, or it may be understood vaguely, as the 'current evidence' or 'what is now generally accepted'. The contextually specified evidential state is not restricted to 'pure data'; it may include relevant theoretical beliefs and also an awareness of some range of alternative proposals. The proposition expressed, according to the rationalist interpretation, by a probability assertion, is that it would be rational, in the given evidential state, to believe the statement in question to the specified degree.

Let me close this section with five residual points.

(1) We have not precluded the possibility of there being a set of functions $d(H/\alpha) = x$ (mentioned above) which determine, for any evidential state α and statement H, the permissible degrees to which one may believe H when in α.

(2) Such a set of permissible d-functions would not be sufficient to determine the rationality of a system of beliefs; we would require, in addition, some way of selecting the evidential components of that system (and this evidential portion could vary depending on the belief whose legitimacy is under scrutiny).

(3) We may define an interval-valued confirmation function $D(H/\alpha) = \{x: d$ is permissible $\land\ d(H/\alpha) = x\}$. This would permit a construal of nonnumerical probability assertions. Thus, 'H is probably true, given α' becomes $D(H/\alpha) = (1/2, 1)$.

(4) Now we may introduce Carnapian c-functions as variant restrictions of d-functions. Thus, $c(H/E) = x$ is a permissible confirmation function if there is a permissible d-function such that for any statements H and E, $c(H/E) = d[H/\alpha(E)]$, where $\alpha(E)$ is the evidential state of being certain of E.

(5) If there is only one permissible c-function, it can be understood to specify real valued logical probability claims. If there are many permissible c-functions, we may define a set-valued C-function: $C(H/E) = [x: c$ is permissible and $c(H/E) = x]$ which may be taken to encode set-valued logical probability statements.

The empirical interpretation

Certain attributions of probability ascribe the tendency of one sort of condition to produce another. They are clearly intended to hold independently of what anyone believes or ought to believe, and for these, any form of subjectivist, rationalist, or logical interpretation would seem to be quite inappropriate. Thus, one may allude to the unknown probability that a particular sort of chemotherapy would cure Hodgkin's disease, or attribute a specific probability to the decay of a radium atom within some designated interval of time. Such empirical probability claims purport to describe objective facts whose discovery requires experimental investigation or some other form of *a posteriori* inquiry. I may believe, and this may be perfectly rational, given the available evidence, to degree 0.9 that Smith will be cured by the chemotherapy – and so, in some sense, 0.9 would be the probability that he will be cured. But this gives no indication of the empirical probability – the tendency of that treatment to produce a cure.

Empirical probabilities will play a minimal role in our subsequent analysis of scientific methodology, and so the following discussion can be skipped by those who are impatient to engage the main topics of this book. My reasons for including this section are: first, to complete the account of probability; second, to preclude confusion by stressing the difference between the empirical brand of probability and those concepts closely affiliated with degrees of belief; and third, to provide an important application of Bayesianism, showing how this approach may explain the confirmation, and therefore the empirical content, of empirical probability hypotheses.

The classical account of such claims is the relative frequency theory.

> The probability that an A (for example a certain kind of experiment or trial) would be an O (for example produce a certain sort of result or outcome) is equal to the limit of the relative frequency of Os in an infinite random sequence of As.

Suppose the existence (in the past, present, and future) of an infinite number of trials of kind A. Consider a particular infinite sequence of such trials, some of which have outcome O:

$$A_1\ A_2\ A_3\ A_4\ A_5\ A_6\ A_7\ A_8\ A_9\ A_{10},\ \ldots, A_n, \ldots$$
$$O\ OO\ \ O\ O\ O$$

Suppose that there is a limit to the relative frequency of Os in this sequence. Then, according to the above definition, this number is the probability that an A-trial will result in outcome O.

However, it is possible to form infinitely many other infinite sequences from this set. Sequences may be formed such that there is no limit to the relative frequency of Os. Other sequences may be formed by carefully selecting their members which are such that the limit is any number between 0 and 1. Thus, it is necessary to say more specifically which sequence of As is to be used for the determination of O's probability.

This is usually done by requiring that the sequence of As be a *random sequence* – its members should be randomly selected from the set of As. One may be tempted to explain this by saying that, for any pair of trials A_k A_j and any position x in the sequence, the probability that A_k fills x is equal to the probability that A_j fills x. But it seems clear that the circularity in this characterisation is fatal.

A second possibility is to characterise a random sequence of As as one whose limit is the same as the limit of a specified class of subsequences. A_1, $A_2, \ldots, A_n, \ldots$ will be random with respect to O only if, for any one-to-one function f (of a certain kind) from the set of positive integers to the set of positive integers, the limit of the relative frequency of outcome O in the sequence $A_{f(n)}$ is the same. Thus, $A_1, A_2, \ldots, A_n, \ldots$ is random with respect to O, only if its limit is the same as for the sequences

$$A_2 A_4 A_6, \ldots$$
$$A_1 A_2 A_4 A_5 A_7 A_8, \ldots$$

Now, we still have no guarantee that every random sequence (in this sense) formed from the set of As has the same limit. In fact, this is clearly false, since a sequence of As consisting solely of those with outcome O, will qualify as random. Clearly, however, the limit of *that* sequence would *not* be regarded as O's probability.

A suspicious feature of this sequence is that many As are left out of it. So it seems natural to use this as a general basis for excluding such undesirable sequences. That is, we may require that the probability that A yield O be derived from a random sequence of As whose members exhaust all of the infinite number of trials of that kind which have ever existed and will exist. Nevertheless, can we be sure that there are not different ways of ordering the whole set of As, which will yield different limits of the relative frequency of O? Again, the answer is no.

At this point it might be well to note that, in practice, probabilities are inferred from sequences generated by the temporal order in which As occur, or by the order in which they are encountered. Thus, one might try to specify the sequence from which O's probability is to be derived, in that sort of way. This seems less than perfectly satisfactory, however, since it introduces elements into the concept of probability which seem intuitively independent of that concept.

Alternatively, we might resign ourselves to doing without a clear characterisation of 'the probability that A yields O,' and rest content with a specification of the idea 'the probability of O in a sequence A_n of As'. This is also unsatisfactory, since probability then becomes a property of an infinite sequence of trials of a certain kind – whereas, in practice, we ascribe probabilities to the kind of trial (and even, arguably, to particular trials).

Let us assume, in order to proceed beyond this first difficulty with the frequentist theory, that the relevant sequence of As is the one engendered by the temporal order in which they occur. Suppose there is such a sequence A_n that it is random with respect to O, and that the limit of O's relative frequency in A_n is x. So, according to the frequentist theory under consideration, it follows that the probability of A yielding O is x. Now, a second difficulty with this theory is that, intuitively, the occurrence of the sequence A_n is quite compatible with the fact that the probability of O is not equal to x. A random sequence of As may have a limit which differs from the actual probability – though the probability of such an occurrence is zero. For example, a perfectly fair coin – one for which the probability of heads is $1/2$ – *may* yield an infinite sequence of tails. There is no contradiction or impossiblity in this supposition – though its probability is zero.

So far we have been assuming that there actually is an infinite set of A-trials, ordered by their time of occurrence to yield a random sequence A_n. But, of course, this assumption is unrealistic. We often refer to the probability that an A-trial would yield outcome O, in cases where we know that there will never be an infinite number of A-trials. We may suspect that *no* A-trials have been or will ever be performed. To make sense of these probability attributions, in the spirit of the orthodox relative frequency interpretation, it is necessary to identify a probability with what the limit *would be* if an infinite series of trials *were* to be performed. But now further difficulties are added to those I have already mentioned. For example, let an A-trial be a tossing of coin x with apparatus y. Consider the claim:

(H_{kn}) If the coin x were tossed n times with apparatus y, the relative frequency of heads would be k_n.

Our modified relative frequency account requires the following:

(1) For each integer n there is a unique number k_n such that (H_{kn}) is true.

(2) The sequence $k_1, k_2, \ldots, k_n, \ldots$ converges to a limit. However, it is implausible to suppose that these requirements will be met.

Firstly, suppose we put $n = 1$. Then the possible values of k_n are 1 or 0. But whichever of these values are substituted in H_{kn} we will obtain a *false* counterfactual. For suppose the probability of heads is really $1/2$. Then if coin x were tossed once with apparatus y, it *might* land heads. And, if coin x were tossed once with apparatus y, it might not land heads. But both the stronger claims

> If coin x were tossed once with apparatus y,
> it *would* land heads
> If coin x were tossed once with apparatus y,
> it *wouldn't* land heads

suggest that in the hypothetical conditions a certain result would be bound to occur. And this is false.

In addition, one may object to the second assumption, that the sequence $k_1, k_2, \ldots, k_n, \ldots$ will converge to a limit; and that this limit is identical to the probability in question. For suppose we put $n = 10^{10}$. In those hypothetical conditions it seems natural to suppose that the coin and apparatus would have worn out (yielding no correct value of k_n) or, at least, that their physical characteristics would have changed, through wear and tear, so substantially that the outcomes would cease to be an accurate reflection of the original probability of getting heads.

In order to escape from these difficulties, and in the context of a general reaction against forms of verificationism, it is natural to abandon the orthodox relative frequency interpretation, and desist from any attempt to give an explicit analysis of empirical probability attributions – without thereby impugning their legitimacy. Now, definitions in terms of observables facilitate the verification of sentences containing the defined terms; and if we have no idea of how one might go about trying to confirm or disconfirm some claim, its significance is indeed suspect. Consequently, with the renunciation of the observational analysis provided by the relative frequency interpretation, it becomes particularly important to give some account of what would qualify as evidence for, and against, attributions of empirical probability.

It is somewhat ironic that, although the relative frequency interpretation owes its life to the positivistic desire for reduction to observables, that very sentiment also gives it the kiss of death. For we can observe only *finite* sequences of trials, and the evidence resulting from such observations – the relative frequency of outcome O – is compatible with any limit of O's relative frequency in an encompassing *infinite* sequence. For any finite sequence, S_f, of trials, and for any x between zero and one, there is an infinite sequence in which S_f is a subsequence, for which the limit of the relative frequency of O is x. A perfectly fair coin may land heads on 1000 consecutive tosses.

Now, from the perspective of positivism this insusceptibility to conclusive verification or falsification renders attributions of probability scientifically vacuous. We reject this extreme view – recognising that scientific claims are normally neither conclusively verifiable nor falsifiable. But something must nevertheless be said about the circumstances in which probability claims are confirmed and disconfirmed.

In practice, the observation of a relative frequency, x, of Os in a large sample of As is regarded as evidence in favour of the hypothesis

H_x: the probability that an A would yield O is x

$$[\Pr(O : A) = x]$$

and the strength of such evidence is believed to increase with the size of the sample. Let us examine the justification for this procedure.

Suppose a coin is picked randomly from a set of which it is known that half are fair and the rest biased in such a way that the probability of heads is $1/10$. The coin is tossed 10 times and lands heads 4 times. The alternative hypotheses under consideration are $H_{1/2}$ (that the coin is fair) and $H_{1/10}$ (that it is biased). We want to see how the evidence $4h$ may provide reason to prefer one of these hypotheses over the other.

According to Bayes' theorem

$$\Pr(X : Y \& Z) = \frac{\Pr(X : Z) \cdot \Pr(Y : X \& Z)}{\Pr(Y : Z)}$$

Now, put H_x for X; $4h$ for Y; and $10A$, plus background information B, for Z.

$$\therefore \ \Pr(H_x : 4h \& 10A \& B)$$
$$= \frac{\Pr(H_x : 10A \& B) \cdot \Pr(4h : 10A \& H_x \& B)}{\Pr(4h : 10A \& B)}$$

B represents background knowledge and includes information about how the coin was selected. Thus

$$Pr(H_{1/2} : 10A \& B) = 1/2$$

and

$$Pr(H_{1/10} : 10A \& B) = 1/2$$

If H_x is true, then the probability that a sequence of n trials of type A will include r instances of outcome O is given by

$$Pr(rO : nA \& H_x) = \frac{n!}{r!(n-r)!} \cdot x^r (1-x)^{n-r}$$

$$\therefore Pr(4h : 10A \& H_x \& B) = \frac{10!}{4!6!} \cdot x^4 (1-x)^7$$

Dividing the two instances of Bayes' theorem, we have

$$\frac{Pr(H_{1/2} : 4h \& 10A \& B)}{Pr(H_{1/10} : 4h \& 10A \& B)}$$

$$= \frac{Pr(H_{1/2} : 10A \& B) \cdot Pr(4h : 10A \& H_{1/2} \& B)}{Pr(H_{1/10} : 10A \& B) \cdot Pr(4h : 10A \& H_{1/10} \& B)}$$

$$= \frac{(1/2)^4 \cdot (1/2)^6}{(1/10)^4 \cdot (9/10)^6}$$

$$\approx 18$$

Thus, the hypothesis that the coin is fair is 18 times more probable than the competing hypothesis that the coin is biased. Moreover, it is easy to see that, given a fixed proportion of heads, say 40%, the longer the sequence which gives that result, the more likely it is that the fair coin hypothesis is correct.

In addition, it can be shown, along the lines of the above argument, that if H_x and H_y are competing hypotheses which are equally likely, relative to our background information, and if our trials have a relative frequency of heads equal to x, then, for any y, H_x is more probable, given the evidence, than H_y. Moreover, as we saw in our examination of swamping of priors (p. 34), the final result is not sensitive to the initial relative probabilities of H_x and H_y.

This sort of procedure for confirming or disconfirming attributions of probability appears to be precisely what we wanted, but it contains an important flaw. In order to reach the conclusion that $H_{1/2}$ should be accepted or rejected, we need some principle which would permit this to be inferred on the basis of the relative probabilities. We have no license so far, given our result

$$\Pr\left(H_{1/2} : 4h \,\&\, 10A \,\&\, B\right), \, > \Pr\left(H_{1/10} : 4h \,\&\, 10A \,\&\, B\right)$$

to conclude that hypothesis $H_{1/2}$ should be preferred – that is, is worthy of a higher degree of belief than $H_{1/10}$ – on the basis of our evidence and background knowledge.

What is required is some principle of inference, linking knowledge of empirical probabilities to appropriate degrees of belief. For example,

> (L) If S knows that the empirical probability that an A would yield an O is x, and knows that k is an instance of A, and has no other relevant information, then S's degree of belief that k will yield O should be x.

Some such principle would enable us to infer from the above result that, given the evidence, one's degree of belief in $H_{1/2}$ should be 18 times greater than one's degree of belief in $H_{1/10}$.

This account of the confirmation of empirical probability statements shows that the very coherence of such statements depends upon the recognition of nonfrequentist, subjective probabilities. In order to assure the empirical content and scientific respectability of 'frequentist' attributions of probability, it is necessary to explain how observations might provide evidence for or against them. But to do this, we must invoke the notion of degrees of belief. And, given the need for (L), we must, moreover, embrace a strong rationalist interpretation of certain probability assertions. It remains to be seen whether (L) is precisely the principle we require; and, also, whether its status is empirical or conventional. These are challenging questions, but their examination goes beyond the scope of this essay. I wanted to show merely that empirical probability is not the only sort of probability we need and to sketch a non-positivistic Bayesian account of its empirical significance.

Let me end with a note concerning future notation. Of the various concepts of probability discussed above, only two will often appear in the following chapters:

(1) *Subjective probability*, which will be formulated as

$$P_S(H) = x$$

meaning, S's degree of belief in H is x. Sometimes the subscript is dropped when it is clear from the context whose degrees of belief are described.

(2) *Logical probability*, which will be formulated as

$$c(H/E) = x$$

meaning, the degree to which E probabilifies H (and, therefore, the degree of belief which anyone should have in H if his total evidence is E) is x.

3

Confirmation

Explications

If your unconditional degree of belief that elephants can fly is lower than your conditional degree of belief relative to the supposition that they have wings, then, at least as far as you are concerned, the information that elephants have wings would count in favour of the view that they are capable of flight. Generalising, we might be inclined towards the following probabilistic account of evidence:

(1) E confirms (supports, is evidence for) H
    iff $P(H/E) > P(H)$
(2) E disconfirms (is evidence against) H
    iff $P(H/E) < P(H)$
(3) E is irrelevant to H
    iff $P(H/E) = P(H)$

But this would be open to a couple of substantial objections. First, it is too subjective. It suggests the conditions in which an individual, given his personal degrees of belief P, would say that E confirms H. But facts about evidence are not simply matters of taste. E may be evidence for H, whatever I may happen to think; and what we want is an account of the circumstances in which this objective fact obtains. Second, as Glymour (1980) has noted, its plausibility is confined to those contexts in which the truth of E is uncertain. When E is known, $P(E) = 1$ and $P(H/E) = P(H)$. Thus, we would have to deny that established data could ever qualify as evidence; and this is obviously wrong.

Nevertheless, this approach is on the right track, and both difficulties can be avoided by the introduction of suitable revisions. To see what is required it is necessary to acknowledge two things. The first is that evidential claims are made relative to background assumptions which are not always explicit. For example, we may suppose that some cloud of smoke is evidence of fire, but only because we think that these phenomena are usually conjoined. The second point is that true evidential claims imply rational constraints upon our beliefs. Given the general knowledge just cited, any reasonable person observing the smoke should be more confident that there is a fire in the vicinity. Therefore, we are led to the following reformulation:

(4) E confirms H relative to B
   iff $R[B \supset P(H/E) > P(H)]$

or, in words, if and only if reason requires that with background assumptions B, anyone's conditional degree of belief in H given E, should be greater than his unconditional degree of belief in H. Similarly, I propose

(5) E disconfirms H relative to B
   iff $R[B \supset P(H/E) < P(H)]$
(6) E is irrelevant to H, relative to B
   iff $R[B \supset P(H/E) = P(H)]$

This account pre-empts both objections. It is no longer subjective, since what is required, on pain of irrationality, is not a matter of individual opinion. For example, whatever may be my degrees of belief in H and in E, we can show that E confirms H relative to the belief state:

$$B : P(H \to E) = 1, 0 < P(H) < 1, 0 < P(E) < 1$$

for reason requires that our beliefs satisfy Bayes' theorem, that is, that

$$P(H/E) = \frac{P(H)P(E/H)}{P(E)}$$

Therefore, given B, $P(E/H)$ should be 1, and so $P(H/E)$ should be greater than $P(H)$.

Moreover, we can now account for confirmation by known facts, whose probabilities equal 1. For it is natural to identify the evidential force of such information with what ought to have been the epistemic effects of its acquisition. In other words, we might say that E, which is known to be true, is evidence for H, by virtue of the fact that, given our beliefs before E was discovered, $P(H/E)$ should have been greater than $P(H)$. Thus, our evidential claims on behalf of established data may be accommodated by taking B to represent the beliefs which constituted our epistemic state prior to the discovery of E – when $P(E)$ did not equal 1.

Along similar lines, we may explicate what it is for E to confirm H more strongly than F does, and what it is for E to confirm H more strongly than it confirms J. At this point, it is important to distinguish between the level of support attained by H with the discovery of E, and the amount of support which would be specifically contributed by E. The former sense of 'degree of confirmation' concerns the total credibility which should attach, all things considered, to H, given the discovery of E; this is indicated by $P(H/E)$. The latter sense of 'degree of confirmation' is captured by the particular contribution of E's discovery to the probability of H. It may be measured either by the ratio, or by the difference there ought to be between $P(H/E)$ and $P(H)$. Thus, we may supplement (4)–(6) with the following definitions:

(7) E confirms H more than F does, relative to B
   iff $R[B \supset P(H/E) > P(H)]$

(8) E *provides* more reason to believe H than J, relative to B
   iff $R[B \supset P(H/E)/P(H) > P(J/E)/P(J)]$

(9) E *leaves* more reason to believe H than J, relative to B
   iff $R[B \supset P(H/E) > P(J/E)]$

Note that when the background assumptions B consist in the belief, to degree one, of a single (conjunctive) statement K, then these explications may be formulated in terms of Carnapian confirmation functions. For example,

(4') E confirms H relative to K
   iff $(c)[c(H/E \wedge K) > c(H/K)]$

where c ranges over all rationally permissible confirmation functions, and $c(q/p)=x$ assigns degree of belief x to q for anyone whose evidential circumstances consist in the certainty of p.

If B includes tentatively held beliefs, this representation is impossible. In that case, we may employ the more comprehensive confirmation functions $d(q/\alpha) = x$ (see pp. 39–42), each of which assigns degree of belief x to q for anyone whose evidential circumstances are α (where α is a set of subjective probability assignments such as: $p \wedge q_{1/2} \wedge r_{1/3} \wedge s \wedge t_{1/4}$). Then we have

(4") E confirms H relative to B
   Iff $(d)[d(H/E_1 \wedge B) > d(H/B)]$

where d ranges over all rationally permissible confirmation functions. Similarly, formulations in these terms may be given of (5)–(9).

The paradox

The so-called paradox of confirmation consists in a conflict between three plausible assumptions:

(N) What is known as Nicod's sufficient condition for confirmation: that any hypothesis of the form 'All ψs are ϕ' is confirmed by statements of the form 'Object k is both ψ are ϕ';

(Q) The equivalence condition: if E confirms H and H is logically equivalent to J, then E confirms J;

(I) A particular intuition: namely, that the observation that something is not black and not a raven should not confirm the hypothesis that all ravens are black.

From (N) we infer that $\overline{B}\overline{R}$ ('k is not black and not a raven') confirms hypothesis J: $(x)(\overline{B}x \supset \overline{R}x)$. But J is logically equivalent to H: $(x)(Rx \supset Bx)$. Therefore, we conclude via (Q) that $\overline{B}\overline{R}$ confirms H, in conflict with (I).

This is a paradox because, although at least one of the crucial assumptions *must* be mistaken, none of them may be easily dismissed. (N) appears to be precisely what we do tacitly assume. (Q) is undeniable: how can we reasonably have more, or less, confidence in one statement than in something we should recognise to be logically equivalent to it. And (I) seems to follow from the fact that ornithology cannot be done from an armchair – you have to go out and look at birds. Nevertheless, despite their initial plausibility, I will argue on the basis of a Bayesian analysis of the problem that (N) and (I) are both false.

In the first place, however, it is worth noting an unrelated and still more bizarre effect of Nicod's criterion, compared with which the familiar 'paradoxes' are quite innocuous. It follows from (N) that, for any property R and pair of objects a and b, the observation that a and b each *lack* R confirms the hypothesis that a *has* R and the hypothesis that b *has* R.

To demonstrate this, consider

(1) $(x)\left[(\overline{R}x \& \overline{R}b) \supset x \neq b\right]$

From (N) we may infer that (1) is confirmed by

(2) $\overline{R}a \& \overline{R}b \& a \neq b$

But (l) is logically equivalent to

(3) Rb

(Instantiate b in (1), and note that (3) invariably falsifies the antecedent in (1).) Therefore, given the equivalence condition, (2) confirms (3). Similarly, the hypothesis

(4) $(x)\left[(\overline{R}x \& \overline{R}a) \supset x \neq a\right]$

is confirmed by (2), and is logically equivalent to

(5) Ra

Therefore, the proposition '$\overline{R}a\,\&\,\overline{R}b\,\&\,a \neq b$' confirms both '$Ra$' and '$Rb$'. An analogous difficulty is presented by

(6) $(x)\left[(Rx\,\&\,Rb\,\&\,\overline{Q}b) \supset Qx\right]$

This should be confirmed by:

(7) $Ra\,\&\,Rb\,\&\,\overline{Q}b\,\&\,Qa$

but (6) is logically equivalent to:

(8) $\overline{R}b \vee Qb$

Therefore (6) is inconsistent with the alleged evidence for it.

Thus, our data may, according to Nicod's criterion, confirm some hypothesis – and yet entail that the hypothesis is false. Hempel (1965) has shown how these difficulties will emerge if Nicod's criterion is extended and applied to universal conditionals in more than one variable. He suggests, however, that the restriction to a single variable would be sufficient to block them. Yet, as we have seen, this is not enough.

Rather, these absurdities may be avoided only by patching up Nicod's criterion – requiring, for example, that the hypotheses to which it may apply contain no individual names and no identity signs. But such a manoeuvre is palpably *ad hoc*, and so it should not be surprising that further difficulties erupt in the form of the traditional paradox of confirmation.

I will not attempt to survey the myriad 'solutions' and vast literature which have been inspired by this topic. My aim here is to expound a particular Bayesian analysis of the problem, and so I shall confine discussion of alternative strategies to a previous Bayesian proposal by J. L. Mackie (1963). That approach is in the tradition of the probabilistic accounts of Hosiasson-Lindenbaum (1940), Alexander (1958), Good (1960) and Suppes (1966). It is, I believe, along the right lines, and the solution it produces is elegant and enticing. But on close scrutiny a number of objectionable features become evident: these will be avoided in the strategy which I shall defend.

The central idea of Bayesian accounts is that our background assumptions concerning the proportion of ravens and black objects in the universe affect the extent to which hypotheses are confirmed by various kinds of evidence. Suppose we believe that the proportion of things which are ravens is very small: call it x; and the proportion of black things y. Then our relevant background assumptions may be represented by the following table:

Table 1

	R	$\overline{R}$	
B	xy	$(1-x)y$	y
$\overline{B}$	$x(1-y)$	$(1-x)(1-y)$	$1-y$

Thus we suppose that the subjective probability of observing a black raven $P(BR)$, is xy; and similarly, $P(B\overline{R}) = (1-x)y, P(\overline{B}R) = x(1-y)$ and $P(\overline{B}\,\overline{R}) = (1-x)(1-y)$.

Now, consider the table which, according to Mackie, would represent the further supposition – All ravens are black:

Table 2

	R	$\overline{R}$
B	x	$y-x$
$\overline{B}$		$1-y$

If H is true, there are no nonblack ravens. Consequently, the entire raven population x must be confined to the BR box; the entire nonblack population must go into the $\overline{B}R$ box; and this leaves a proportion of $1 - x - (1-y) = y - x$ for the $B\overline{R}$ box. In other words, $P(BR/H) = x, P(B\overline{R}/H) = y - x, P(\overline{B}R/H) = 0,$ and $P(\overline{B}\,\overline{R}/H) = 1 - y$.

Now Mackie employs what he calls the inverse principle, whereby the strength of evidence E in favour of hypothesis T is given by $P(E/T)/P(E)$ – the factor by which the assumption of the hypothesis increases the likelihood of the evidence. But

$$\frac{P(BR/H)}{P(BR)} = \frac{1}{y}$$

and

$$\frac{P(\overline{B}\,\overline{R}/H)}{P(\overline{B}\overline{R})} = \frac{1}{1-x}$$

and

$$x \approx 0$$

So he obtains the nice result that although the observation of a nonblack nonraven *will* tend to confirm 'All ravens are black', it will do so only to a negligible degree and will not carry as much weight as the observation of a black raven.

Thus, of the various possible reactions to the paradox – reject (N) or (E) or (I) – Mackie opts to abandon (I). And he can both justify this response and, moreover, explain why we were tempted to hold the mistaken intuition: quite understandably we failed to distinguish negligible confirmation from no confirmation at all. His solution would benefit from some defence of the inverse principle, but this, as he recognises in a later article (1969), can be readily provided. By a rearrangement of Bayes' theorem:

$$\frac{P(E/H)}{P(E)} = \frac{P(H/E)}{P(H)}$$

Therefore, what Mackie takes to measure evidential value is indeed the factor by which the probability of H would be increased by the discovery of E.

Neat as it is, Mackie's strategy involves a number of disconcerting features. In the first place, it turns out that

$$\frac{P(B\overline{R}/H)}{P(B\overline{R})} = \frac{y - x}{(1 - x)y} < 1$$

Therefore, allegedly, the observation of a black nonraven should not, as we might think, be irrelevant, but should slightly disconfirm the hypothesis. This does not, of course, constitute a decisive objection, since we have already allowed that intuition is prone to error. If we may confuse slight with no confirmation in the case of nonblack nonravens, then we should not be surprised by an additional failure to distinguish irrelevance from slight disconfirmation. Nevertheless, this result is a minor embarrassment.

Second, the information which constitutes our evidence is very sensitive to the way in which the data are gathered. It is one thing to pick out an object at random and discover it to be a black raven; it is quite another thing to take a known raven (selected, say, by a special machine which has been programmed to bring one back), examine its colour, and find out that it is black; and it is yet a third thing to establish that some machine-selected black thing is a raven. These alternative methods of observing a black raven engender different items of evidence (respectively, BR, R^*B, and B^*R) which are not guaranteed to confirm our hypothesis to the same degree. On the contrary, we would expect that the best evidence would be provided when a machine-selected raven turns out to be black ($R*B$), for in that case, H is subject to the maximum risk of falsification and has passed the most severe test. But when a known black thing is scrutinised and found to be a raven ($B*R$), this evidence does not test the hypothesis at all: no discovery about it could jeopardise the hypothesis.

However, Mackie's account is at odds with these intuitions. Employing his tables we find

$$\frac{P(R^*B/H)}{P(R^*B)} = \frac{\text{proportion of } R \text{ which is } RB \text{ in Table 2}}{\text{proportion of } R \text{ which is } RB \text{ in Table 1}}$$

$$= \frac{1}{y}$$

and

$$\frac{P(B^*R/H)}{P(B^*R)} = \frac{x/y}{x} = \frac{1}{y}$$

Thus, we are told that equal support is provided by all three items of evidence. Again, this is implausible.

Finally, and this is the source of the previous difficulties, there is no foundation for Mackie's crucial presupposition about our background assumptions – that our views concerning the proportions of ravens and black things are independent of whether or not we are supposing that H is true. In other words, it is essential to Mackie's solution that

$$P(B/H) = P(B)$$

and

$$P(R/H) = P(R)$$

But these claims have no prima facie plausibility and they lead directly to the problems just described.

To see their importance, remember that it is only by holding x fixed as we move to the H table that we force that increase in the supposed population of the BR box, which leads to the significant confirmation provided by BR. Similarly, keeping y constant requires the smaller proportional increase in the supposed population of the $\overline{B}\overline{R}$ box. And these changes entail a decrease in the supposed population of the $B\overline{R}$ box, which yields the counterintuitive result that $B\overline{R}$ disconfirms H.

So the three flaws in Mackie's solution are not independent of one another. The idea that the subjective probability of selecting a raven, and the probability of selecting a black thing, are independent of whether or not we are supposing that all ravens are black, plays a fundamental role in his solution but also produces the counterintuitive results. Thus, it is worthwhile to consider whether a Bayesian approach, minus this troublesome and unjustified assumption, can yield a satisfactory solution.

I shall initially assume nothing about the prior and conditional background beliefs, except that they are coherent, that is, conform to the probability calculus.

So let

$$
\begin{array}{ll}
P(BR) = a & P(BR/H) = \alpha \\
P(B\overline{R}) = b & P(B\overline{R}/H) = \beta \\
P(\overline{B}R) = c & P(\overline{B}R/H) = 0 \\
P(\overline{B}\,\overline{R}) = d & P(\overline{B}\,\overline{R}/H) = \delta
\end{array}
$$

as shown in the following tables.

Table 3

	R	$\overline{R}$
B	a	b
$\overline{B}$	c	d
	Background	

Table 4

	R	$\overline{R}$
B	α	β
$\overline{B}$		δ
	Background +H	

Applying Bayes' theorem we have

$$
P(H/BR) = \frac{P(H)P(BR/H)}{P(BR)} = \frac{\alpha}{a} \cdot P(H)
$$

$$
P(H/\overline{B}\,\overline{R}) = \frac{P(H)P(\overline{B}\,\overline{R}/H)}{P(\overline{B}\,\overline{R})} = \frac{\delta}{d} \cdot P(H)
$$

However, there is no reason to suppose that

$$
\frac{\alpha}{a} > 1 \ \text{ or } \ \frac{\delta}{d} > 1 \ \text{ or } \ \frac{\alpha}{a} > \frac{\delta}{d}
$$

and, consequently, no reason to conclude that

$$P(H/BR) > P(H) \text{ or } P(H/\overline{B}\,\overline{R}) > P(H)$$

or

$$P(H/BR) > P(H/\overline{B}\,\overline{R})$$

In other words, if we observe a random object, neither the discovery that it is a black raven, nor that it is a nonblack nonraven count in favour of 'All ravens are black'.

This may look unpromising; but there is still room to manoeuvre. As I said above, the observation of a black raven may take place in various ways. It may consist of discovering the colour of something known to be a raven, or finding out that some known black thing is a raven, or just happening to come across an object which turns out to be a black raven. Similarly, there are various ways of gathering the data that something is a nonblack nonraven. I have argued that there is no reason to suppose that happening upon a black raven provides more support for the hypothesis than coming across a nonblack nonraven, or indeed, that either provides any support. But it remains to be seen how the various alternative forms of evidence perform. Consulting the tables we find

$$P(H/R^*B) = \frac{P(H)P(R^*B/H)}{P(R^*B)} = \frac{P(H).1}{a/(a+c)} = \frac{a+c}{a} \cdot P(H)$$

$$P(H/B^*R) = \frac{P(H)P(B^*R/H)}{P(B^*R)} = \frac{\alpha}{a} \cdot \frac{(a+b)}{(\alpha+\beta)} \cdot P(H)$$

$$P\left(H/\overline{B}^*\overline{R}\right) = \frac{P(H)P\left(\overline{B}^*\overline{R}/H\right)}{P\left(\overline{B}^*\overline{R}\right)} = \frac{d+c}{d} \cdot P(H)$$

$$P\left(H/\overline{R}^*\overline{B}\right) = \frac{P(H)P\left(\overline{R}^*\overline{B}/H\right)}{P\left(\overline{R}^*\overline{B}\right)} = \frac{\delta}{d} \cdot \frac{(d+b)}{(\delta+\beta)} \cdot P(H)$$

$$P(H/B^*\overline{R}) = \frac{P(H)P\left(B^*\overline{R}^*/H\right)}{P(B^*\overline{R})} = \frac{\beta}{b} \cdot \frac{(a+b)}{(\alpha+\beta)} \cdot P(H)$$

$$P\left(H/\overline{R}^*B\right) = \frac{P(H)P\left(\overline{R}^*B/H\right)}{P\left(\overline{R}^*B\right)} = \frac{\beta}{b} \cdot \frac{(b+d)}{(\beta+\delta)} \cdot P(H)$$

Now, there is nothing to rationally dictate values of a, b, c, d, α, β, and δ which would make either $P(H/B^*R)$ or $P\left(H/\overline{R}^*\overline{B}\right)$ diverge from $P(H)$. The same goes for $P(H/B^*R)$ and $P\left(H/\overline{R}^*B\right)$; thus we avoid one of Mackie's counterintuitive results. However, if we suppose, which is eminently plausible, that it is part of our background knowledge that the proportion of nonblack nonravens among things in the universe is nearly 1 – that is to say, $d \approx 1$ – then, $1/a \gg 1/d > 0$ therefore $1 + c/a \gg 1 + c/d > 1$; so we may infer

$$P(H/R^*B) \gg P\left(H/\overline{B}^*\overline{R}\right) \overset{\text{slightly}}{>} P(H)$$

This relationship between our degrees of belief is required by reason, relative to the background assumption that $d \approx 1$. Thus, given that assumption, the discovery that a known raven is black substantially confirms H; and the discovery that a known nonblack thing is not a raven negligibly confirms it.

This, I think, solves the paradox of confirmation. We have found:

(1) Nicod's criterion is false – not only, as we might have initially thought, because of what it has to say about nonblack nonravens – but more significantly because it entails that the random observation of positive instances must confirm a hypothesis. But this is mistaken in general, and in particular when it is reasonable to put $a > \alpha$. For in those circumstances, the observation of a positive instance of a hypothesis will disconfirm it. A nice illustration is given by Swinburne (1971):

All grasshoppers are located in regions other than Pitcairn Island

Swinburne points out that, contrary to the dictates of Nicod's criterion, the observation of a grasshopper off Pitcairn Island (and especially the observation of *many* such positive instances) might rather be expected to increase our confidence that grasshoppers are numerous and widespread, and thus reduce our confidence in the truth of the hypothesis. This intuition is explained by the fact that in such a case it is reasonable to put $a > \alpha$. Assuming the absence of grasshoppers on Pitcairn Island would reasonably indicate a smaller grasshopper population than would otherwise be supposed, reducing the subjective probability that a positive instance would be observed.

(2) There is something right about the idea that a hypothesis of the form $(x)(\psi x \supset \phi x)$ is confirmed by the observation of positive instances. The grain of truth is that the discovery that a known ψ is ϕ must qualify as favourable evidence. But it is another matter entirely to suppose, in accordance with Nicod's criterion, that an object selected at random, and found to be both ψ and ϕ, conveys support. As we have seen, that is not generally true. We must beware of a tendency to confuse the evidential value of distinct items of information which may each be taken to constitute positive instances of a hypothesis.

(3) The intuition (I), concerning the irrelevance of nonblack nonravens is capable of various interpretations. Construed as a claim about $\overline{B}\overline{R}$ or $\overline{R}^*\overline{B}$, the intuition is correct – those items of data neither confirm nor disconfirm. Understood as a claim about $\overline{B}^*\overline{R}$, the intuition is, strictly speaking, false; $\overline{B}^*\overline{R}$ confirms, though the support it provides is negligible.

(4) Although it has been shown that R^*B – and to a lesser extent $\overline{B}^*\overline{R}$ – confirms the general hypothesis 'All ravens are black', we have not seen how these items of data bear upon particular predictions. In other words, we have been given no reason to conclude that the evidence R^*B should confirm the belief that the *next* raven to be examined will be black. It might be thought that from

$$R^*B \text{ confirms } H$$

and

$$H \text{ entails } R^{\text{next}}B$$

we could conclude

$$R^*B \text{ confirms } R^{\text{next}}B$$

Such reasoning employs what is known as the consequence condition:

$$\text{If } E \text{ confirms } H \text{ and } H \text{ entails } J \text{ then } E \text{ confirms } J.$$

However, we can see very quickly that this principle is false. Suppose $P(B/A) < P(B)$, and let H be $A \wedge B$, E be A and J be B. Then E confirms H (since $P(A \wedge B/A) > P(A \wedge B)$), clearly H entails J, but E does not confirm J. Thus, we can't invoke the so-called consequence condition to justify our projection of positive instances of 'All ravens are black'. It remains for us to examine what actually does justify that inference: this is the problem of induction. In the next section I shall explore a natural, yet misguided, attempt to solve it. Since the discussion is somewhat technical and, after all, a dead end, it could well be skipped or skimmed.

A Bayesian pseudo-solution to the problem of induction

Russell's formulation (1912) of the principles of inductive reasoning is reminiscent of Nicod:

When a thing of a certain sort A has been found to be associated with a thing of a certain sort B and has never been found dissociated from a thing of the sort B, the greater the number of cases in which A and B have been associated, (a) the greater is the probability that they will be associated in a fresh case in which one of them is known to be present; (b) the more probable it is that A is always associated with B; and a sufficient number of cases of association will make these probabilities approach certainty without limit.

If H_1 is the hypothesis, 'All As are B', and 'nAB' is taken to represent the statement 'n known As have been found to be B' then Russell's formulation may be summarised as follows: in advance of the acquisition of any data, a rational assignment of subjective probabilities P must satisfy:

(1) $P(H_1/(n + 1)AB) < P(H_1/nAB)$
(2) $P((n + 1)AB/nAB) > P(nAB/(n - 1)AB)$
(3) $\text{Limit}_{n \to \infty} P(H_1/nAB) = 1$
(4) $\text{Limit}_{n \to \infty} P((n + 1)AB/nAB) = 1$

It is instructive to consider how a Bayesian justification of these principles might proceed. The probability calculus yields the following constraints:

$$P(H/E) = \frac{P(H)P(E/H)}{P(E)}$$

$$P(E) = P(H)P(E/H) + P(H')P(E/H') + P\left(H''\right)P\left(E/H''\right) + \cdots$$

where $H, H', H'', \ldots$ are a set of exhaustive and mutually exclusive hypotheses.

$$\therefore P(H/E) = \frac{P(H)P(E/H)}{P(H)P(E/H) + P(H')P(E/H') + \cdots}$$

Now, let H_j stand for the hypothesis:

The proportion of A s which are B is j

Then our initial subjective probability assignments should satisfy

$$P(H_j/nAB) = \frac{P(H_j)P(nAB/H_j)}{P(nAB)}$$

Expanding $P(nAB)$ in terms of those mutually exclusive H_j to which we give non-zero credibility, we obtain

$$P(H_j/nAB) = \frac{P(H_j)P(nAB/H_j)}{\Sigma_j P(H_j)P(nAB/H_j)}$$

where Σ_j designates summation of terms for all those values of j such that $P(H_j) \neq 0$. But if one knew only that j is the proportion of As which are B, then j should be the degree of belief that some particular known A is B. This is eminently plausible, although it does go beyond the constraints of the probability calculus. That is,

$$P(1AB/H_j) \text{ should be } j$$

and

$$P(nAB/H_j) \text{ should be } j^n$$

So, for a rational subject

$$P(H_j/nAB) = \frac{P(H_j)j^n}{\Sigma_j P(H_j)j^n}$$

$$\therefore P(H_1/nAB) = \frac{P(H_1)}{\Sigma_j P(H_j)j^n}$$

Now, as $n \to \infty$

$$j^n \to 0 \text{ for } 0 \leq j < 1$$

and

$$j_n = 1 \text{ for } j = 1$$
$$\therefore \Sigma_j P(H_j)j^n \to P(H_1)$$
$$\therefore P(H_1/nAB) \to 1$$

establishing (3): the credibility of the hypothesis approaches one as the number of favourable observations tends to infinity.

In addition, it is a theorem of the probability calculus that if X entails Y, and not vice versa, then the probability of X is less than the probability of Y. But H_1 entails that the $(n + 1)$th A will be B

$$\therefore P(H_1/nAB) < P((n+1)AB/nAB)$$

So, given the previous result, as $n \to \infty$

$$P((n+1)AB/nAB) \to 1$$

establishing (4).

Now to prove (1), simply note that

$$(n+1)AB \text{ entails } nAB$$
$$\therefore P(H_1/(n+1)AB) = P(H_1/(n+1)AB \wedge nAB)$$

Applying Bayes' theorem to the right-hand side

$$P(H_1/(n+1)AB) = \frac{P(H_1/nAB) \cdot P((n+1)AB/H_1 \wedge nAB)}{P((n+1)AB/nAB)}$$

but

$$P((n+1)AB/H_1 \wedge nAB) = 1$$

and

$$P((n+1)AB/nAB) < 1$$
$$\therefore \ P(H_1/(n+1)AB) > P(H_1/nAB) \qquad \text{QED}$$

Finally, to prove (2),

$$\frac{P((n+1)AB/nAB)}{p(nAB/(n-1)AB)} = \frac{\dfrac{P((n+1)AB) \cdot P(nAB/(n+1)AB)}{P(nAB)}}{\dfrac{P(nAB) \cdot P((n-1)AB/nAB)}{P((n-1)AB)}}$$

$$= \frac{P((n+1)AB) \cdot P((n-1)AB)}{[P(nAB)]^2}$$

$$= \frac{\left(\Sigma_j P(H_j) j^{n+1}\right)\left(\Sigma_j P(H_j) j^{n-1}\right)}{\left(\Sigma_j P(H_j) j^n\right)^2}$$

$$= \frac{\Sigma_j P(H_j)^2 j^{2n} + \Sigma_{(j,k)} P(H_j) P(H_k)(jk)^{n-1}\left(j^2 + k^2\right)}{\Sigma_j P(H_j)^2 j^{2n} + \Sigma_{(j,k)} P(H_j) P(H_k)(jk)^{n-1}(2jk)}$$

When $j = k$, then $j^2 + k^2 = 2jk$

When $j \neq k$, then $(j - k)^2 > 0$

that is,

$$j^2 + k^2 - 2jk > 0$$

that is,

$$j^2 + k^2 > 2jk$$

Therefore, some terms in the numerator are greater, and none smaller, than corresponding terms in the denominator.

$$\therefore \frac{P((n+1)AB/nAB)}{P((nAB)/(n-1)AB)} > 1$$

Thus, the probability that the next A will be B increases as the number of favourable observations increases.

All of Russell's principles of induction may appear to be justified by these Bayesian arguments. However, the existence of some mistakes in the reasoning is strongly suggested by consideration of the grue problem. Let 'grue' mean 'sampled and green or unsampled and blue'. Then it would seem that, according to Russell's principles, our accumulating data concerning the greenness (and therefore grueness) of sampled emeralds should eventually provide us with arbitrarily good reason to believe both that all emeralds are green and that they are grue, and justify inconsistent expectations about those emeralds which have not yet been sampled.

In general, let J be a hypothesis of the form

$$\text{All } A \text{ s are } C$$

where

$$Cx \overset{\text{definition}}{\equiv} (Bx \& Sx) \vee (\overline{B}x \& \overline{S}x)$$

and

$$\text{'}Sx\text{' means '}x \text{ is sampled'}$$

If we are in possession of the data represented by nAB, we are also in a position to know nAC. Therefore, the above principles of induction should permit the same observations to confirm to an arbitrarily high degree both '$(n + 1)AB$' and '$(n + 1)AC$'. But these expectations may be satisfied only if the next (as yet unsampled) A is both B and $\overline{B}$, respectively. Since the expectations are incompatible, the sum of their subjective probabilities should not exceed one; and they cannot both be arbitrarily well confirmed. Thus, it appears that the Russellian principles of induction must be incorrect, and their Bayesian rationale invalid.

However, there is an alternative way to resolve the problem. We may locate the source of confusion in a failure to recognise that our evidence is not simply nAB, as has been assumed, but rather $nASB$. That is to say, the data from which we wish to generalise consist in the information that n *sampled As* were found to be B. But now, the relevance of principles (1)–(4) is thrown into doubt. For they concern probabilities relative to information of the form nAB; whereas what we should be interested in are $P(H_1/nASB)$ and $P((n + 1)AB/nASB)$. Moreover, the original principles are not inconsistent. nAB and nAC are not equivalent. So it is perfectly coherent to suppose that, for sufficiently high n, the data nAB establish H_1 and the data nAC establish J.

In order to regenerate the problem we would have to adopt revised principles such as

$$(3') \ \underset{n\to\infty}{\text{Limit}}\, P(H_1/nASB) = 1$$

This would certainly cause trouble. For ASB is equivalent to ASC; and from (3') we could obtain

$$\underset{n\to\infty}{\text{Limit}}\, P(J/nASC) = 1$$

Thus (3') runs afoul of the grue problem, and it is important to see why it cannot be justified in the manner of (3).

Paralleling the argument for (3), we may say

$$P(H_j/nASB) = \frac{P(H_j)P(nASB/H_j)}{\Sigma_j P(H_j)P(nASB/H_j)}$$

But whereas in the earlier argument we had to assume

$$P(AB/H_j) = j$$

and

$$P(nAB/H_j) = j^n$$

which are quite plausible, we would in this case have to suppose

$$P(ASB/H_j) = j$$

and

$$P(nASB/H_j) = j^n$$

But these cannot be generally correct. For they embody the idea that our sample is unbiased – that the chances of an A turning out to be B are unaffected by its presence in the sample. And whilst this may be maintained in certain cases, it cannot be held in general. If being in S does not affect the (subjective) chances that an A is B, then it *will* affect the chances that it is $SB \vee \overline{S}\,\overline{B}$ (that is, C) – as long as it is not supposed that there are an equal number of B and $\overline{B}$. Therefore, if we can assume $P(ASB/H_j) = j$, we cannot also assume $P(ASC/J_k) = k$. In other words, for some j,

$$P(AB/H_j) \neq P(A\overline{B}/H_j)$$

So, if a sample is not biased with respect to B,

$$P(AB/H_j) = P(ASB/H_j)$$

and

$$P(A\overline{B}/H_j) = P(A\overline{S}\,\overline{B}/H_j)$$
$$\therefore\ P(ASB/H_j) \neq P(A\overline{S}\,\overline{B}/H_j)$$
$$\therefore\ P(ASC/H_j) \neq P(A\overline{S}C/H_j)$$

The sample *is* biased with respect to C. Thus we cannot suppose, in general, that

$$P(ASB/H_j) = j$$

The principle (3′) cannot therefore be justified; and this is just as well, given the inconsistencies which would emerge in the face of 'grueified' hypotheses.

Projection

We have seen how it follows from Bayes' theorem that hypotheses are confirmed by discovering the truth of their logical consequences. In particular, both

All emeralds are green

and

All emeralds are grue

are confirmed, their probabilities augmented, by the earlier observation of green emeralds. But it is extremely important not to confuse this result with claims about the evidence regarding unobserved cases. It is one thing to maintain, as I have, that

$$H = \text{All As are } B$$

is supported – its probability raised – by evidence of the form $nASB$. But it is entirely another matter to suppose that such evidence makes it more likely that the unsampled As are also B. In other words, we have

$$P(H/nASB) > P(H)$$

but we do not have

$$P((n+1)AB/nASB) > P(nAB/(n-1)ASB)$$

In the latter case, when the observation of positive and no negative instances tends to confirm the conclusion that the unsampled As are B, the hypothesis H is said to be *projectible*.

In what circumstances is H projectible? Is there some condition on the prior probability of H, whose satisfaction is necessary and sufficient for H to be projectible? We can make some progress with these questions.

Let
a represent :	all sampled As are B	
b represent :	all unsampled As are B	
H represent :	$a \wedge b$; and $P(H) = x$	
J represent :	$a \wedge \bar{b}$; and $P(J) = y$	
K represent :	$\bar{a} \wedge b$; and $P(K) = z$	
L represent :	$\bar{a} \wedge \bar{b}$; and $P(L) = w$	

H is projectible iff:

$$P(\text{the unsampled } As \text{ are } B/\text{the sampled } As \text{ are } B)$$
$$> P(\text{the unsampled } As \text{ are } B)$$

that is,

$$P(b/a) > P(b) \equiv \frac{x}{x+y} > x + z$$
$$\equiv x^2 + (y + z - 1)x + yz < 0$$
$$\equiv x^2 - (x + w)x + yz < 0 \quad (\text{since } x + y + z + w = 1)$$
$$\equiv xw > yz$$

This entails that $yz < \frac{1}{16}$. For suppose, on the contrary, that $yz \geq \frac{1}{16}$. Any pair of numbers α and β must satisfy the relation $\alpha\beta \leq [(\alpha + \beta)/2]^2$. Therefore,

$$\left(\frac{y+z}{2}\right)^2 \geq \frac{1}{16}$$
$$\therefore \ y + z \geq \tfrac{1}{2}$$

but

$$x + y + z + w = 1$$
$$\therefore \ x + w \leq \tfrac{1}{2}$$
$$\therefore \ \left(\frac{x+w}{2}\right)^2 \leq \frac{1}{16}$$
$$\therefore \ xw < \tfrac{1}{16}$$
$$\therefore \ xw \leq yz$$

Thus, a necessary condition for H to be projectible is that $P(J)P(K)$ (the product of the prior probabilities of the grue-like alternatives to H) be less than 1/16. In addition, we can see that projectibility is associated with high prior probability. In most cases $P(L) = w \approx 1$, and then, for H to be projectible, it is sufficient that $x > y$ or $x > z$.

These results are admittedly quite meagre. They are first attempts to express the idea of projectibility in probabilistic terms. I have not said anything about the characteristics of a hypothesis, by virtue of which we assign it a particular prior probability sufficient for projectibility. Nor have I yet discussed the rational ground for such assignments. Nevertheless, some understanding of the notion of projectibility has been gained.

We are also now in a better position to evaluate the highly controversial consequence condition:

If E confirms H and H entails R, then E confirms R

In the first place, this principle is clearly inconsistent with our criterion of confirmation. E may raise the probability of a hypothesis and yet diminish the probability of some consequences of the hypotheses. This was shown a few pages ago: simply let s and q be statements such that $P(q/s) < P(q)$; and put $H = s \wedge q$, $R = q$, and $E = s$.

Our problem is to reconcile its evident falsity with the fact that the consequence condition appears to be a central element of scientific methodology: we do, it seems, take the evidence for a theory to provide us with reason to believe its predictions.

The solution is simple. There do exist hypotheses, the observation of whose positive instances we feel provides no reason to believe that they will be satisfied in future – all emeralds are grue, for example. In those cases the hypotheses themselves are confirmed, but claims about particular unobserved instances, which follow from the hypotheses, are disconfirmed. But the characteristic which distinguishes these unprojectible hypotheses is a relatively low prior probability. We have seen that hypotheses with relatively high prior probabilities are projectible and thus do satisfy the consequence condition. And these, not surprisingly, are the hypotheses which are typically proposed by scientists, offered as explanations of the data. Consequently, it is to be expected that the consequence condition would characterise scientific practice, even though it is false in general.

What has emerged in the last two sections is that my Russellian formulation of the principles of induction is inadequate. As they stood initially, they were justifiable in Bayesian terms yet inapplicable to any actual evidential circumstances; but when this deficiency is rectified, they fall foul of the grue problem and become inconsistent. Only so-called projectible hypotheses conform to Russell's principles. We have made some progress with an attempt to reduce projectibility to probabilistic concepts, and may conclude that the projectibility of a hypothesis derives from its relatively high prior probability. Finally, we have seen why the consequence condition may be widely employed, even though it is false.

4

Induction

The traditional problem of induction derives from Hume's question: 'What is the nature of that evidence which assures us of any real existence and matter of fact beyond the present testimony of our senses or the records of memory?' His answer is that we expect the future to resemble the past and 'expect similar effects from causes which are to appearance similar'. But he argues that this assumption cannot be deductively justified (since no contradiction arises from its denial), and cannot be established by reasoning appropriate to matters of fact (since such reasoning would require the very assumption to be justified, and would therefore be circular). And so he infers that 'our conclusions concerning matters of fact are not founded on reasoning or any process of the understanding'. In short, Hume's view is that these beliefs are determined by the unjustifiable instinctive expectation that nature will be uniform.

As we shall see, there has been little improvement upon Hume's line of thought, although these days the discussion is more refined. In particular, it now seems clear that the problem of induction – to investigate the rational basis of inductive inference – cannot be resolved without considerable preliminary attention to (A), the nature of inductive inference; and to (B), what should be required in a justification of it. Thus, there are three distinct questions here, and three respects in which an alleged solution may be open to criticism. Firstly, it may involve, and exploit, an erroneous conception of our inductive practice. As the grue problem reveals, we do not generally accept the rule

All the many sampled As have been B
∴ Probably, all As are B

Secondly, it could presuppose mistaken standards of justification. Obviously, we should not be required to show that inductive arguments are invariably truth preserving. And thirdly, it may fail, even within its own terms, to establish that what is taken to constitute our inductive reasoning does meet the supposed adequacy conditions.

In this chapter I will discuss each facet of the problem. To begin, I consider the general nature of those constraints which yield conditions for a system of beliefs to be rational, and I conclude that Carnapian confirmation functions provide a good way to represent them. The question then arises as to what must be done to justify our conformity to certain c-functions rather than others. But the answer is remarkably difficult. A natural idea is to require *noncircular demonstrable reliability*: proof by a non-question-begging argument that the allegedly strong arguments (according to the c-function in question) are frequently truth-preserving. However, we shall find, first, that this requirement cannot be sufficient (if our inductive practice is correct), for there is a well-known confirmation function c^+ which satisfies it, but which obviously does not encode our inductive reasoning. Second, I'll argue that Lewis' *immodesty* constraint cannot be invoked to exclude certain confirmation functions, since they all are immodest. Third, we shall see that what c^+ lacks and our practice has is *audacity* – the tendency to go out on a limb and sanction extreme degrees of belief. And, fourth, it transpires that this very lack of audacity is what is responsible (and necessary) for $c^{+\prime}$ s satisfaction of the requirement; so that, if our practice is reasonable, then noncircular demonstrable reliability cannot even be a necessary condition. Finally, I will devise a weaker (circular) requirement of demonstrable reliability, which is trivially satisfied by our practice. This won't help with the problem of justification; to that end I sketch a form of the so-called semantic justification of induction.

(A) The nature of inductive inference

Modus ponens does not require of someone who accepts p and $p \to q$ that he should set about believing q as well. For he might reasonably elect to abandon p. Rather, its role is to legislate upon the rationality of certain combinations of beliefs. It demands that, ideally, if his beliefs include p and $p \to q$, then q should already be included. Similarly, any *inductive* practice is a disposition to decide, in a particular way, whether certain combinations of beliefs are permissible or irrational. In other words, it consists in the implicit demand that particular constraints be met by any complex of beliefs which is to qualify as rational. One such condition, generally acknowledged, is that an epistemic state should be consistent. Its elements should not contradict one another; nor should they entail incompatible consequences. Another constraint, for which I have argued, is that the constituent beliefs be coherent: conform to the principles of elementary probability theory.

It seems to me an open question whether or not rationality demands the satisfaction of further constraints, though I tend to think that it does. I suggested this in Chapter 2, criticising the account of probability which I called 'weak rationalism'. For it is difficult to see how merely the requirements of consistency and coherence can yield such results as:

(a) 'All emeralds are green' is worthy of a higher degree of belief than 'All emeralds are grue'.
(b) It is unlikely that the accepted principles of physics are accurate only until tomorrow morning at which time the material universe will cease to exist.

The point here is not simply that (a) and (b) may be denied without self-contradiction, or that their denials may be consistently embedded within some more comprehensive sets of beliefs. For there may well be circumstances in which such bizarre epistemic states are perfectly rational. It is rather that a person could coherently embed the denials of (a) or (b) within a system of beliefs, which, in other respects, is substantially similar to our own. And, in that case, the deviant claims would be regarded as irrational despite the absence of inconsistency or incoherence.

If there are extra constraints, what might they look like? I think it is useful to consider this question in terms of the typology of constraints, shown in Fig. 8.

What I mean by an internal constraint is one whose application depends only upon the intrinsic content of an epistemic state. Consistency is an internal constraint; for in order to determine whether or not it is satisfied,

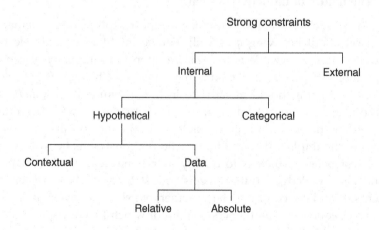

Fig. 8

it suffices to know simply which beliefs are contained in the epistemic state. One need not know, for example, how the beliefs were caused, or anything about the previous history of the subject. An external constraint, on the other hand, implies that the rationality of a system of beliefs depends on circumstances which go beyond the identity of the beliefs. It has been suggested, for example, that the rationality of certain kinds of belief may depend upon their generation by a reliable belief-producing mechanism – a process, such as visual perception, which tends to result in true beliefs – regardless of the subject's awareness of this fact.

It seems to me that the postulation of such extrinsic constraints is misguided. For rationality is a normative concept; the adoption of beliefs in the absence of justifiability is wrong and subject to disapprobation. But such an attitude would be quite inappropriate if the conditions for justification included circumstances whose presence we could not be expected to recognise. The normative character of justification requires that there be no divergence between what we would be justified in believing and what we would be justified in thinking we would be justified in believing; and this substantially constrains the conditions for justification and rational belief. In particular, it precludes such a condition as *produced by a reliable mechanism*. For it is perfectly conceivable that a method for acquiring beliefs be absolutely reliable, even though we have every reason to distrust it. We cannot *prove* that every otherwise plausible external constraint must run afoul of this condition. However, it is clear that internal constraints are more likely to satisfy it. Insofar as the rationality of a system of beliefs depends simply upon its having properties whose possession may be inferred from the identity of the beliefs, then we can reasonably expect the subject, who is aware of his beliefs, to determine whether or not they are rational. But if extrinsic constraints govern justification, especially if aspects of the external environment become pertinent, then there arises the possibility of a reasonable, yet incorrect, estimate of the presence of those conditions, and this may yield the paradoxical divergence between what I ought to believe and what I ought to believe I ought to believe.

Within the class of internal constraints I distinguish two kinds: *categorical* and *hypothetical*. Consistency and coherence are examples of categorical constraints. Their application does not involve any division of the system – any discrimination between evidence and theory, premises and conclusions. Hypothetical constraints, on the other hand, determine which conclusions ought to be accepted, and to what degree, on condition that certain other things are assumed. Their application engenders a disposition to evaluate in a particular way the credibility of statements in a range of evidential circumstances. Thus, a hypothetical constraint may be represented by a function $D(H/\alpha) = I$, which specifies, for certain statements H and evidential circumstances α, a set I of the degrees of belief which it would be rational to attach to H in the evidential circumstances α. It is important to see that the mere acceptance of such a function does not suffice to determine the rationality of an epistemic state, and therefore does not define an inductive practice. Its use for that purpose depends upon a further decision to take some portion of the combination of beliefs as evidence whose legitimacy is not called into question and to which the remaining beliefs must be responsible in a manner prescribed by the function. Thus, a hypothetical constraint has two components: a function D, and a way of specifying the evidential portion of an epistemic state.

Assuming (what has not been established) that internal constraints over and above consistency and coherence comprise our conception of rationality, it is a further open question whether such additional constraints are categorical or hypothetical. This is a substantial matter since it cannot be assumed that categorical constraints may, in general, be cast into the form of hypothetical ones. So, if it is taken for granted (as it was by Carnap, for example) that rationality is a matter of having appropriate theoretical beliefs, given the evidential circumstances, there is a danger that any accurate characterisation of our inductive practice will already have been precluded.

In fact, Carnap's presuppositions are even stronger – and in two respects. In the first place, he assumes that a person's evidential circumstances may be identified with his acceptance of a single conjunctive observation sentence – one which records his entire observational knowledge. Thus, he becomes confined to confirmation functions of the form $c(q/p) = x$ which specify for any statements p and q the strength of an argument from conjoined premises p to conclusion q. And this move is difficult to reconcile with two plausible possibilities (see p. 39) (a) that our evidence may contain uncertain, tentative beliefs, and (b) that the credibility of some explanatory hypothesis may be determined, in part, by the context of explicitly recognised alternative theories. In rejecting the relevance of our horizon of alternatives, Carnap's c-functions qualify as *data* constraints. In the second place, Carnap assumes that the rationality of any system of beliefs is verified by substituting for p the conjunction of all data statements in the system, and then checking to make sure that all other statements are believed to the degree prescribed by some permissible c-function. The data statements are unrevisable. More and more will accumulate with further observation, but none, once accepted, may ever

be renounced. This does not square well with a third and a fourth plausible possibility: (c) that theoretical considerations may cast doubt upon
what was believed on the basis of observation, and (d) that within a
system of beliefs the portion which constitutes the evidential state varies,
depending upon which particular belief is in question. The typology
includes a category for *relative constraints* in order to accommodate this
possibility. They are distinguished from the set of *absolute* constraints
implicit in Carnap's approach, whereby the evidential state is independent
of which belief in the system is to be justified.

Thus, well in advance of any difficulties which may derive from more
specific claims about precisely which c-functions are permissible, Carnap's
model of our inductive practice involves six questionable assumptions:
(1) that not every consistent and coherent system of beliefs is rational; that
the further conditions are (2) intrinsic, (3) hypothetical, and (4) data
constraints, (5) that evidential circumstances consist in the acceptance of
a proposition, and (6) that this proposition is the conjunction of what is
unrevisably known on the basis of observation.

It is worth repeating that the existence of strong c-functions requires
only assumptions (1), (2) and (3), which are the most plausible of the six.
The others become involved through Carnap's view about the way in
which these functions should represent our inductive practice and determine the rationality of epistemic states. Now, I have stated and defended
my inclination to accept (1) and (2). Moreover, I shall accept (3) for the
following reason. In scientific and everyday contexts it is plainly the
practice to support hypotheses by citing further beliefs as evidence, and
it is our practice to estimate the quality of such arguments – to assess the
plausibility of their conclusions in the light of those considerations
brought forward in defence of them. Thus, it is reasonable to suppose that
our inductive practice may be represented by a function which specifies,
for any evidential circumstance, the permissible degrees of belief in any
statement.

I have noted that our so-called 'evidential circumstances' may include
not only the possession of tentative data, but also an awareness of some
spectrum of alternative theories. And these points give rise to the question
whether c-functions could possibly play a significant role in confirmation
theory. The answer, I think, is yes; and it seems to me that there are two
ways to defend it.

One way is to proceed along the lines indicated in Chapter 2, p. 39 to
show, first, that uncertain observation can be accommodated by means of
Jeffrey's rule; and second, that the awareness of a context of alternative

theories may be treated as one more evidential belief, to be plugged in to the confirmation function in conjunction with data beliefs. If this can be done, then an inductive practice may be totally characterised by means of a c-function, together with a way of identifying evidential states.

Second, suppose the problems of uncertain data and contextual awareness cannot be settled along these lines. In that case, we can restrict our focus to those somewhat idealised situations in which we are *not* hampered or influenced by an inability to formulate theories and *are* aware of all the relevant alternatives. The restricted permissible confirmation functions for such situations would be what I have labelled d-functions, from an uncertain observational belief state $p_{1/2}$, $q_{1/4}$, $r_{1/3}$, ... to degrees of belief in hypotheses. And now, in terms of this function, one may define an even more restricted c-function. Thus, even if the hypothetical constraint of our inductive practice cannot be fully captured by a set of c-functions because our practice recognises such a broad class of evidential constraints, nevertheless our practice is *partially* characterised by such a set. Consequently, in discussing the justification of our practice we may reasonably begin by concentrating on the question: what is the rational basis for our conformity to those c-functions which characterise (either partly or wholly) our inductive practice?

To end this section let me relate these conclusions to the earlier discussion of projectibility. We have seen that Russell's characterisation of our inductive practice is incorrect. We do not accept, without crucial qualifications, that nature is uniform. In other words, we do not suppose, in general, that the chances that all As are B, and that the next A is B, are increased to virtual certainty by a sufficiently large sample of positive instances. More precisely, our practice is not represented by any confirmation function c, such that for *any* A and B

$$c(\text{All } As \text{ are } B / nASB) \rightarrow 1 \quad \text{as } n \rightarrow \infty$$

or

$$c((n+1)AB/nASB) > c(nAB/(n-1)ASB)$$

However, there are cases in which this sort of reasoning is taken to be sound, and indeed to constitute paradigm examples of good inductive inference (for example when 'A' means 'emerald' and 'B' means 'green'). Part of the problem of describing our inductive practice – what Goodman has called 'the new riddle of induction' – is to provide a specification of those non-grue-like predicates for which Russell's account is accurate. I have argued that this riddle reduces to that of describing the basis of our prior probability assignments. For it has been shown that the project-ibility of 'All emeralds are green' derives from its high prior probability. Note that the notion of prior probability is relative. A probability may qualify as prior relative to certain data (for example emerald observations) but not relative to other information (concerning laws of physics, for example). Thus, what has been shown is that, given background infor-mation K such that

$$c(\text{All } As \text{ are } B/K) \text{ is fairly high}$$

then

$$c(\text{All } As \text{ are } B/nASB \wedge K) \text{ becomes very high as } n \to \infty$$

and

$$c((n+1)AB/nASB \wedge K) > c(nAB/(n-1)ASB \wedge K)$$

The two problems of induction remain. First, the so-called 'new riddle', which is to find a systematic characterisation of those combinations of A, B and K, such that the antecedent is true. Or, in other words, to find a function c which conforms to our intuitive judgement about the value of

$$c \text{ (All } As \text{ are } B/K)$$

And second, the traditional problem: to justify our conformity to whatever confirmation function does in fact capture our intuitions; in other words, to justify those intuitions.

(B) Conditions of rationality

We take it for granted that our inductive practice is reasonable: conformity is encouraged, and any departure from it is deprecated as irrational. The problem of induction consists in the critical examination of this assump-tion, and it is clear that such a project cannot proceed without careful attention to what the reasonableness or unreasonableness of our practice would depend upon. What must be done to justify induction?

Demonstrable reliability
I shall consider a variety of adequacy criteria which have been proposed in the literature on this subject. The first, and most prevalent, view is that a reasonable inductive practice must be *demonstrably reliable*: there must be reason to think that it is good at directing us towards the truth. For

profound and true beliefs are our objective, and induction is our way of attempting to achieve it. So it can be reasonable to adopt one practice rather than another only if it can be shown that the one is more likely to be successful.

More specifically, we would want to show that supposedly stronger inductive arguments (with true premises) have true conclusions more frequently than arguments which, according to our practice, are weaker. Consider some evidential circumstance a, relative to which certain independent hypotheses are each deemed to warrant degree of belief 0.9, and certain other hypotheses are each ascribed in our practice a probability of 0.7. The practice will have directed us properly only if more of the first set of hypotheses are true than of the second set. Moreover, it will have got things exactly right if 90% of the first, and 70% of the second set of hypotheses turn out to be true. In general, if S_x^a is the class of statements whose members are each assigned credibility x relative to evidential conditions a, then the relative frequency RF, with which members are true should be approximately x. So we may state a requirement of demonstrable reliability as follows:

(DR) An inductive practice represented by the confirmation function $c(p/E) = x$ is justified only if it may be deductively proved from independently established premises that, for any true statement E

$$\mathrm{RF}(p \text{ is true}/p \in S_x^E) = x$$

where

$$S_x^E = \{p : c(p/E) = x\}$$

I have argued above that our inductive practice is at least partially representable by a set of Carnapian c-functions. In my formulation (DR), and in the following discussion, I concentrate on just that part of our practice. This simplification permits a particularly clear presentation of problems and proposals in the area. And it is a legitimate strategy, given the fact that any justification of the whole practice would have to involve a justification of that part of it.

(DR) provides one, not wholly implausible, example of a requirement of demonstrable reliability. This sort of requirement is often proposed, at least informally: see, for example, the work of Brian Skyrms (1975) and Michael Friedman (1978). But I am not claiming that (DR) is correct. On the contrary, there are alternative formulations, less stringent than (DR), and I will argue later that one of these is the right condition to impose. For example:

(a) We may relax the demand that $\mathrm{RF}(p$ is true$/p \in S_x^E) = x$ and require only approximate equality.
(b) We may drop the insistence upon a noncircular demonstration and allow some inductive rationale for the correspondence between credibility and relative frequency.
(c) We may require merely that the credibilities prescribed by c determine an expected value (in the technical sense) equal to x, for the proportion of truths in S_x^E.

Whichever requirement of demonstrable reliability provides a necessary condition of reasonableness, we cannot assume that it is also sufficient. For all we know, more than one possible practice may satisfy the condition. Or it may be that although only one meets this requirement, this practice is inadequate on other grounds – failing to satisfy some further necessary condition of reasonableness which we have yet to identify.

The demonstrable reliability of $c^\dagger$

These are realistic concerns, as the following argument will confirm. For there is a possible inductive practice, obviously not our own, which clearly satisfies (DR): namely, that practice represented by the confirmation function

$$c^\dagger(p/E) = \text{the proportion of those possible worlds}$$
$$\text{in which } E \text{ is true, where } p \text{ is also true}$$
$$= \frac{\text{number of } (pE)\text{-worlds}}{\text{number of } E\text{-worlds}} = \frac{k_{pE}}{k_E}$$

where k_q is the number of worlds where q is true. This confirmation function was advocated in Wittgenstein's *Tractatus* (5.15) and examined by Carnap in *Logical foundations of probability*. Clearly its application is restricted to languages in which only a finite number of possible worlds (N) may be distinguished – one world for each complete state description in the language. It is defined for a language with three names 'a', 'b' and 'c', and one monadic predicate 'F' In this case, '$Fa \wedge - Fb \wedge Fc$', is a typical complete state description; eight possible worlds – one for each complete state description – can be distinguished; and, for example,

$$c^+(Fa/-Fa \lor Fb) = \frac{k_{Fa \,\&\, (-Fa \lor Fb)}}{k_{-Fa \lor Fb}}$$
$$= \frac{k_{Fa \,\&\, Fb}}{k_{-Fa \lor Fb}}$$
$$= \frac{2}{6} = \frac{1}{3}$$

Similarly c^+ is defined for more powerful languages, provided their resources are confined to a finite stock of names, predicates and relations, so that only a finite number of state descriptions and possible worlds may be distinguished.

The following argument demonstrates that c^+ meets the condition of demonstrable reliability (DR): we will prove, *a priori*, by noncircular reasoning, that someone whose inductive reasoning is in conformity with c^+ will in fact achieve exactly the appropriate rate of success – the frequency with which conclusions are true will match the degrees of confidence which are prescribed by c^+. Consider arbitrary statements A and B, where B is true.

$$c^+(A/B) = \frac{k_{AB}}{k_B}$$

Let $Q_B(A)$ be the number of propositions which have the same degree of confirmation, given evidence B, as A does. And let $Q_B^*(A)$ be the number of these propositions which are true. So what we must prove is that

$$\frac{Q_B^*(A)}{Q_B(A)} = c^+(A/B) = \frac{k_{AB}}{k_B}$$

or, in other words, that there is a correspondence between degree of confirmation and relative frequency of truth.

To calculate the magnitude of $Q_B(A)$ note that

$$Q_B(A) = \text{the number of propositions whose degree}$$
$$\text{of confirmation is the same as } As$$
$$= \text{the number of propositions } \phi \text{ such that}$$
$$c^+(\phi/B) = c^+(A/B)$$

that is,

$$\frac{Tk_{\phi B}}{k_B} = \frac{k_{AB}}{k_B}$$

that is,

$$k_{\phi B} = k_{AB}$$

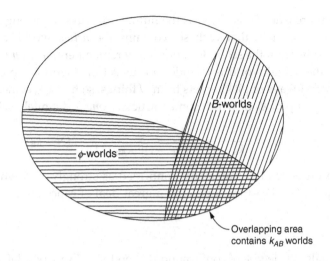

Fig. 9

(as shown in Fig. 9). Thus $Q_B(A)$ is the number of propositions which are true in exactly k_{AB} of the worlds in which B is true. Now let us assume that there is a one-to-one mapping between propositions and sets of possible worlds: that for each statement there is a particular set of possible worlds in which it is true, and for each such set there is a single proposition which is true precisely in those worlds (but which may be expressed by a variety of logically equivalent sentences). In that case we can calculate $Q_B(A)$ by calculating how many sets of possible worlds there are which share k_{AB} elements with the set of B-worlds. Then $Q_B(A)$ is the number of ways of picking a set of possible worlds from the set of all N possible worlds, so that k_{AB} of them come from a particular group of size k_B. Each set is made up of two parts – the part selected from the k_B B-worlds, and containing k_{AB} elements, and second, the part from the remaining $N-k_B$ worlds in which B is not true. Therefore,

$$Q_B(A) = \text{(the number of ways of picking a set of}$$
$$k_{AB} \text{ things from a set of } k_B \text{ things)}$$
$$\times \text{ (the number of ways of picking a subset}$$
$$\text{(any size) from a set of } N - k_B \text{ things)}$$
$$= \frac{k_B!}{(k_B - k_{AB})!k_{AB}!} \cdot 2^{(N-k_B)}$$

Now $Q_B^*(A)$ is calculated similarly. It is the number of statements which, not only (as before) have the same degree of confirmation as A, but which in addition are true. Thus, $Q_B^*(A)$ is the number of ways of picking a set of possible worlds such that each set contains the *actual* world, and k_{AB} elements come from the set of B-worlds. Now remember that: *ex hypothesis*, B is true: the actual world is included in the set of B-worlds. So $Q_B^*(A)$ is the number of ways of picking sets from N things, so that k_{AB} come from a particular group of k_B of which some particular *one* is definitely selected

$$\therefore Q_B^*(A) = \frac{(k_B - 1)!}{[(k_B - 1) - (k_{AB} - 1)]!(k_{AB} - 1)!} \cdot 2(N - k_B)$$

Now, the proportion of statements confirmed as strongly as A which are true – $Q_B^*(A)/Q_B(A)$ – turns out, after a great deal of cancelling, to be

$$\frac{k_{AB}}{k_B}$$

which equals A's degree of confirmation. Therefore, $c^\dagger$ satisfies (DR): it has noncircular demonstrable reliability.

However, as Carnap recognised, there are substantial differences between our inductive practice and what is prescribed by $c^\dagger$. Most significantly, we suppose some properties are projectible – that the attribution of certain properties becomes more strongly confirmed as our evidence comes to include more and more previous cases of instantiation. In other words, as n increases, Fa_1 & Fa_2 &...& Fa_n provide stronger support for Fa_{n+1} But this vital feature of our practice is not captured by $c^\dagger$. Consider the set of worlds in which 'Fa_1 & ...& Fa_n' is true. If F is independent of the other predicates, then, for every such world in which a_{n+1} is also F, there is a corresponding world in which a_{n+1} is not F. Therefore, a_{n+1} is F in half of the worlds in which Fa_1 &...& Fa_n is true. Thus,

$$c^\dagger(Fa_{n+1}/(Fa_1 \& \ldots \& Fa_n)) = 1/2$$

regardless of n. Thus, $c^\dagger$ fails to capture the way in which we learn from experience.

If a demonstration that degrees of confirmation correspond to relative frequencies is necessary *and sufficient* for an inductive practice to be justified, then our practice is irrational. For $c^\dagger$ does satisfy that condition, and ours is inconsistent with $c^\dagger$; so we should be using $c^\dagger$. Therefore, assuming that our practice is correct, the identity of credibility and probability cannot be what is responsible. Its legitimacy must derive from some idiosyncratic characteristic, yet to be identified. Until this feature is made explicit, it will remain unclear what must be accomplished in a justification of our practice; and the problem of induction cannot be well defined.

Immodesty

A further adequacy condition has been suggested by David Lewis: the requirement of immodesty. Any inductive practice generates, in light of the current data, estimates of the values of quantities – the masses and velocities of objects, the heights of people, etc. But the *reliability* of an inductive practice may be quantified, and so every inductive practice yields an estimate of the reliability of each inductive practice, including itself. The requirement of immodesty is that no reasonable inductive practice should estimate that some alternative would be more reliable than itself.

This indeed appears to be a just requirement. However, I shall argue that it cuts no ice since *every* inductive practice is immodest.

The expected value of a magnitude x with possible values $x_1, x_2, \ldots$ is defined as

$$\sum_k x_k \text{Prob}\,(x_k)$$

Thus, according to an inductive practice represented by confirmation function c, the expected value of x, given evidence E, is

$$\sum_k x_k c(x_k/E)$$

First, I shall show that every inductive practice is immodest with respect to each particular estimation that it makes. Suppose our evidence is E and that our estimate of the value of x is λ. The actual inaccuracy or degree of unreliability of this estimate may be measured by

$$(x^* - \lambda)^2$$

where x^* is the actual value of x. According to inductive practice c, the expected inaccuracy of this estimate is

$$M = \sum_k (x_k - \lambda)^2 c(x_k/E)$$

$$\therefore \frac{dM}{d\lambda} = -2\sum_k (x_k - \lambda)c(x_k/E)$$

$$= -2\left[\sum_k x_k c(x_k/E) - \lambda\sum_k c(x_k/E)\right]$$

$$= -2\left[\sum_k x_k c(x_k/E) - \lambda\right]$$

But M is at a minimum when $dM/d\lambda = 0$, that is, when

$$\lambda = \sum_k x_k c(x_k/E)$$
$$= \text{the expected value of } x \text{ according to } c.$$

Thus, if we estimate the expected value, the estimates yielded by c must have, according to c, the greatest expected accuracy. Every inductive practice is immodest with respect to its reliability in particular estimating tasks.

A similar argument demonstrates that they are also immodest about *overall* reliability. Let $x, y, z, \ldots$ be the magnitudes to be estimated; $x_1, x_2, \ldots,$ $y_1, y_2, \ldots, z_1, z_2, \ldots$ their possible values; and $x^*, y^*, z^*, \ldots$ their actual values. Let $e, f, g,$ be the sequences of evidential circumstances; $e_1, e_2, \ldots, f_1, f_2, \ldots,$ $g_1, g_2, \ldots$ their possible values; and $e^*, f^*, g^*, \ldots$ their actual values. (In other words, there are many possible sequences of evidential states, for example $e_1 + f_3 + g_2 + \ldots$, or $e_5 + f_1 + g_3 + \ldots$, etc.) Let $\lambda(x, e)$ be our estimate of x based on e. The square error of this estimate is

$$[x^* - \lambda(x, e)]^2$$

Suppose we make every possible estimate: in each individual circumstance in which we find ourselves, we estimate the values of all the magnitudes. Then our overall inaccuracy may be measured by the sum of the square errors of all the estimates which we make. This overall inaccuracy equals

$$[x^* - \lambda(x, e^*)]^2 + [y^* - \lambda(y, e^*)]^2 + \ldots$$
$$+ [x^* - \lambda(x, f^*)]^2 + [y^* - \lambda(y, f^*)]^2 + \ldots$$
$$\vdots \qquad\qquad \vdots$$

The expected value of this total inaccuracy I, according to c, equals the sum of each possible value, weighted by the probability (given our present evidence E) of obtaining it. We must consider every possible combination of values of $x, y, z, \ldots, e, f, g, \ldots$, multiply the probability of each such combination by the overall inaccuracy which it would determine, and add all these terms together. Thus,

$$I = \sum_{\langle j,k,\ldots l,m,\ldots \rangle} c\big(x_j \wedge y_k, \ldots, e_l \wedge f_{m'} \ldots / E\big)$$
$$\times [x_j - \lambda(x, e_l)]^2 + [y_k - \lambda(y, e_l)]^2 + \ldots$$
$$+ [x_j - \lambda(x, f_m)]^2 + [y_k - \lambda(y, f_m)]^2 + \ldots$$
$$\vdots \qquad\qquad \vdots$$

where $j, k, \ldots, l, m, \ldots$ range over positive integers.

$$\therefore \frac{\partial I}{\partial \lambda(x, e_2)} = -2 \sum_{\langle j, k, \ldots, m, \ldots \rangle} c(x_j \wedge y_k, \ldots, e_2 \wedge f_m, \ldots / E) \cdot [x_i - \lambda(x, e_2)]$$

$$= -2 \left[\sum_{\langle j, k, \ldots, m, \ldots \rangle} x_j c(x_j \wedge y_k, \ldots e_2 \wedge f_m, \ldots / E) \right.$$

$$\left. - \lambda(x, e_2) \sum_{\langle j, k, \ldots, m, \ldots \rangle} c(x_j \wedge y_k, \ldots, e_2 \wedge f_m, \ldots / E) \right]$$

$$= -2 \left[\sum_j x_j c(x_j \wedge e_2 / E) - \lambda(x, e_2) c(e_2 / E) \right]$$

But I is at a minimum value with respect to the estimate $\lambda(x, e_2)$ when $\partial I / \partial \lambda(x, e_2) = 0$. That is, when

$$\sum_j x_j c(x_j \wedge e_2 / E) = \lambda(x, e_2) c(e_2 / E)$$

that is,

$$\lambda(x, e_2) = \sum_j x_j c(x_j / e_2 \wedge E)$$
$$= \text{the expected value of } x, \text{ according to } c$$

Thus, according to c, expected overall inaccuracy is minimised by estimating that magnitudes will take on those expected values determined by c.

The strategy above differs substantially from previous approaches of Lewis (1971, 1973) and Spielman (1972), though our results are similar: they conclude that at least all of Carnap's λ-methods (a special class of c-functions) are immodest. Unlike theirs, my argument (a) is not restricted to the λ-methods, (b) involves no approximation and (c) takes into account the probability that we will find ourselves in various evidential circumstances and will therefore be confronted by various estimating tasks.

Audacity

We have seen that neither demonstrable reliability, nor immodesty, nor their conjunction is sufficient to guarantee that an inductive practice is reasonable. What else might be required? A plausible candidate is *audacity*. We want to know definitely, one way or the other, whether statements are true or false; and if such certainty cannot be obtained we would like to be almost certain. In other words, we prefer a practice which has a greater tendency to sanction extreme degrees of belief – near 1 and 0. Thus, it may be that both c_1 and c_2 satisfy the requirement of demonstrable reliability – in both cases their credibility assignments to hypotheses may be shown to correspond to the frequency with which those hypotheses are true. Nevertheless, c_1 is preferable because it is more audacious: in most evidential circumstances, more statements are assigned high and low credibilities by c_1 than by c_2.

Unlike the other necessary conditions which have been discussed, the requirement of audacity is a matter of degree – the more audacious the better. It remains to give a precise characterisation of the notion, and this will be done to some extent in the next section, when the need arises. However, I think the requirement is intuitively clear and plausible enough to sustain the following provisional conclusion: an inductive practice is justified if and only if it is consistent, coherent, immodest, and the most audacious practice which meets the requirement of demonstrable reliability.

(C) The justification of induction

I have tentatively suggested that, to justify our inductive practice, it is necessary and sufficient to show that it has five characteristics:

(1) Consistency: I shall simply assume that our practice does embody this requirement.
(2) Coherence: the legitimacy of this requirement was demonstrated in the section on subjectivism in Chapter 2. Again, I shall simply assume that we do in fact feel constrained by it.
(3) Immodesty: I have argued that every coherent practice is immodest.
(4) Demonstrable reliability.
(5) The maximum audacity compatible with (1), (2), (3), and (4).

A crucial question concerning (4) is, what is to count as *demonstrability*? What would qualify as a *reason* for supposing that our practice is reliable? On this question there are three schools of thought.

According to the first and, initially, most tempting view, there is inductive evidence for the reliability of our practice. It has been reliable so far and therefore should continue to be trusted. This notion was rejected by Hume as viciously circular, and most philosophers have agreed with him.

Second, there is the idea that, regardless of our empirical data, our practice may be employed to provide reasons to believe in its reliability. This strategy appears to be similarly vitiated by circularity, assuming the very thing that we want to prove.

Third, and most stringent, there is (DR). Some purely deductive rationale is demanded: something which does not presuppose the legitimacy of
our practice – something analogous to our demonstration of the reliability
of $c^\dagger$. This is generally thought to be impossible, and since the only
alternatives are circular and therefore illegitimate, it is concluded that
the problem of induction is insoluble: Hume was right and our inductive
practice is unreasonable.

The impossibility of a noncircular rationale

Let me begin with the third line of thought. Clearly, a deductive rationale
would be best if we could get it. Why should we suppose that this cannot
be done? After all, $c^\dagger$ has been validated deductively! I am not suggesting
that a deductive justification of our practice is in fact on the cards, but only
that the possibility should not be dismissed as quickly and scornfully as it
often is in the literature. It will not do, for example, simply to maintain that
this would turn induction into deduction, or that the future course of
events is not entailed by our present evidence. What is required is the
identification of the relevant difference between our practice and the one
described by $c^\dagger$, together with a demonstration that it is by virtue of this
particular feature that no purely deductive, noncircular argument for the
reliability of our practice can be provided.

I have claimed that what is unsatisfactory about $c^\dagger$, compared with our
practice, is its lack of audacity. I will now argue that it is that very
deficiency which permits the conclusive demonstration of its reliability,
and it is because our practice is more audacious that it cannot be justified
in this way. I want to prove that no inductive practice which is more
audacious than $c^\dagger$ can satisfy (DR).

First, remember that a practice is audacious to the extent that it dictates
extreme degrees of belief. This may be explicated as follows: c_1 is more
audacious than c_2 only if there is some possible evidential circumstance
relative to which, for every x ($>1/2$), the number of statements assigned
credibilities more than x is at least as great for c_1 as for c_2, and for some x it
is greater.

Second,

$$c^\dagger(q/p) = \frac{k_{pq}}{k_p}$$

where k_p is the number of possible worlds in which p is true. Let the
possible values of $c^\dagger(q/B)$ be $x_1, x_2, \ldots$ in order of magnitude. Then x_1 is

the credibility of each of the statements which are true in all the worlds in which B is true; x_2 attaches to the statements which are true in all but one of the B-worlds; and, in general, x_k attaches to statements which are true in all but $(k-1)$ of the B-worlds.

Third, suppose c represents some practice which is necessarily reliable, and more audacious than $c^\dagger$. There must be some smallest positive integer n, such that c prescribes credibilities (given B) less than x_{n-1} and greater than or equal to x_n to a set S of statements which differs from the set of statements to which $c^\dagger$ prescribes such credibilities. S cannot contain any statement which is true in more than (k_B-n) B-worlds, for all such statements have been given credibilities greater than or equal to x_{n-1}. Suppose we remove from S each statement p, which is true in fewer than (k_B-n) B-worlds, and replace each such p with every statement which is true in every Bp-world and also true in exactly (k_B-n) B-worlds. Then each such manoeuvre increases, for certain B-worlds, the proportion of statements in S which are true in those worlds; and for no B-world does the proportion of truths in S ever decrease. Thus, we generate a set of statements S^*, whose proportional reliability in any B-world never drops below x_n, and which is a subset of the statements true in (k_B-n) B-worlds. If it were a proper subset, then there would be B-worlds in which the proportion of S^* truths would fall below x_n. Therefore S^* is the set of all the statements true in (k_B-n) B-worlds, and its proportional reliability in every B-world is x_n, Consequently, $S = S^*$ contradicting our initial supposition. Therefore, no inductive practice is necessarily reliable and yet more audacious than $c^\dagger$: if any practice satisfies (DR), it is not more audacious than $c^\dagger$.

Inductive demonstration of reliability

If the requirement of demonstrable reliability is captured by (DR), then our inductive practice is not demonstrably reliable, and we must conclude, as many philosophers have done, that inductive inference is unreasonable and that beliefs based upon it are unjustified. There is, however, an alternative strategy, a second construal of demonstrability. According to this view, the requirement of demonstrable reliability is satisfied by an inductive practice, provided that *any* good reasons exist for supposing that it is reliable. It is assumed, moreover, that inductive reasons are good reasons. Of course, given such an assumption, any concern with conditions of reasonability such as (DR) becomes irrelevant to the central issue, since the assumption is the very thing whose truth was in question. But a requirement of demonstrable reliability may nevertheless be maintained. By permitting the employment of our inductive practice in the demonstration of its reliability, we do not undermine the legitimacy of the requirement, but only its importance; for we cannot think that such a demonstration would provide support for our practice.

Let us therefore postpone the question of justification, and consider how reliability may be demonstrated with the use of our inductive practice. According to one school of thought, mentioned at the beginning of this section, there is empirical evidence for the reliability of our practice: namely, its past record. Our practice has been successful, so we have reason to be confident about it. If, on the other hand, it had persistently failed us, then we would be justified in expecting yet more inaccuracy in the future and in ceasing to conform our beliefs to its requirements.

However, this is incoherent. Suppose the evidence of past unreliability is E. Our practice C specifies credibilities for every evidential state – including those which contain knowledge of E. Suppose that it dictates

$$C(A/B) = r$$

and

$$C(A/B \wedge E) = s \neq r$$

and that we conform our degrees of belief accordingly. This surely does not involve an abandonment of the practice: on the contrary, it is merely a further application of it. The only way in which it can come about that evidence does justify some degree of belief, in conflict with the requirements of some inductive practice, is if we are not reasoning in accordance with that practice.

I am not denying that we might employ a method M for arriving at beliefs, whose reliability is confirmed or disconfirmed by empirical evidence. I am saying that, in such cases, M could not constitute our inductive practice. Thus, M may be the rule: Whenever smoke is observed, infer the existence of some fire in the vicinity. But the very fact that further information will influence the reasonability of this rule, shows that it is merely *derived* from our inductive practice, which really has the form

$$C(\text{There is a fire near } p/\text{There is smoke at } p \wedge B) = x$$

where x depends upon background assumptions B. And these commitments are insensitive to evidence.

Let me reiterate these points in the context of Max Black's reasoning (1958). He argues as follows for the view that there can be inductive support for the inductive method, R = the rule of inference:

> x out of y instances of A have been B
>
> ∴ The probability is x/y that the next A will be B.

Observation (1). p out of q instances of F have been G ($p/q > 1/2$).

Observation (2). s out of t applications of R have been successful ($s/t > p/q$) (*an application of R is a case in which x out of y instances of A have been B, and $x/y > 1/2$, and it is predicted on the basis of this information that the next A will be B; a successful application of R is one in which the prediction is true*).

Conclusion (1′). The probability is p/q that the next F is G.

Conclusion (2′). The probability is s/t that the next application of R is successful.

(3) The next application of R is known to be our inference from (2)–(4), and the prediction that the next F is G.

∴ (4) (From (2′) and (3)) The probability is s/t that the next F is G.

∴ (5) The recognition of 'second order' information concerning the success rate of previous applications of R permits us to assert those conclusions sanctioned by R with a greater degree of confidence (s/t) than would otherwise have been justified (p/q).

∴ (6) In some sense, R supports itself if it is successful.

Which is the fallacious move in this argument depends upon how we construe R. If it is taken to imply that the conclusion should be inferred from the premise no matter what else is known, then it is self-contradictory. For the evidence provided by (1), (2) and (3) will legislate more than one probability for the claim: the next F is G. If, on the other hand, it is understood merely to specify what should be concluded when our total evidence consists in knowledge of the premise, then the move to (4) is fallacious. For, according to the rule under this construal (1′) follows if one's evidence is (1) alone, and (2′) follows if one's evidence is (2) alone. But the rule R does not tell us what we ought to believe in circumstances where we know (1), (2) and (3). We can certainly further elaborate upon our inductive practice, and specify it in such a way that (4) is justified on the basis of (1), (2) and (3), and in such a way that statement (5) is true. But that would be to correct the idea that R constitutes our inductive practice, and to abandon the concluding moral – R does not support itself, nor does the elaborated inductive method.

So much for the idea of empirical inductive evidence for the reliability of our practice. Let me now turn to the third possibility – the third way of understanding the requirement of *demonstrable* reliability. This is the idea that our practice may be legitimately employed in demonstrating its reliability. Not via some inductive argument from empirical premises; but rather through some *a priori* reasoning in which the commitments of our practice may be presupposed. As I have said, such a thing could not be regarded as a *justification* of induction. That would have to come from elsewhere; and indeed it may, as we shall see presently when we consider the so-called 'semantic' justification of induction. However, this possibility does not invalidate the requirement of demonstrable reliability. It remains intuitively legitimate and so it is worth thinking about how we could show that our practice meets the requirement, if we were to know, and could assume in the argument, that it is reasonable. The question is, what form of the requirement would permit us to show that our practice satisfies it? And one answer is as follows: that there be some *inductive* reason to expect that supposedly good arguments have true conclusions most of the time. More specifically, we might require that the credibilities prescribed by our practice determine *an expected value* equal to x for the proportion of truths in certain sets of statements whose members are assigned credibility x. More precisely,

> (DR2) Let $c(q/a) = x$ represent *our* inductive practice. Any practice represented by the confirmation function $k(q/a) = i$ is justified only if c determines that, for any T_x^{ka} the expected value of the proportion of truths in T_x^{ka} is x: where $s \in T_x^{ka}$ iff $k\,(s/a) = x$

This is a very weak version of the requirement of demonstrable reliability. However, it is worth seeing that there is at least some version which is compatible with the correctness of our inductive practice. I won't go through the argument, since it is quite straightforward: its conclusion is an instance of the theorem due to De Finetti (1937), and discussed by Jeffrey (1982): that the expected proportion of truths in a set of statements is equal to their average probability.

A further benefit of (DR2) is that it allows us to show that our practice will satisfy my fifth condition of reasonability: that there is no more audacious practice which meets the other four conditions – consistency, coherence, immodesty, and demonstrable reliability. For no practice other than our own may satisfy (DR2). Thus, assuming (DR2), our inductive practice satisfies the necessary and sufficient requirements of reasonability which I listed at the beginning of this section, and is therefore correct.

However, (DR2) is a lot to assume since it obviously presupposes the legitimacy of our inductive practice. If what we want is a non-question-begging argument in favour of our practice, the above considerations will not suffice.

Semantic justification

We must look to the so-called 'semantic' justification of induction. This is the idea, associated particularly with Strawson (1952), that the statements

Given evidence E, one ought to believe H to degree x

which characterise our practice, are analytic. So that to ask whether they are really true, or whether our practice is really justified is like asking whether bachelors are unmarried. From this viewpoint it is true by definition of 'rational' that a belief is rational just in case it is formed in accordance with our inductive practice.

I won't attempt to defend this theory, at least directly. Instead, I would like to sketch an alternative semantic argument for the correctness of our inductive practice: one which does not employ the concept of analyticity.

I assume that the *use* of language determines reference, in the following sense: consider a possible world W whose physical and mental characteristics are exactly like ours except for one difference; whenever the expression α appears in our world, whether written, spoken, or thought, the expression β appears in the same location in W. Then it cannot be that some utterance in W containing β is true in W, yet false here, when β is replaced by α (unless that utterance is about language). So, for example, if α is the word 'dog' and β the term 'yog', then we would be inclined to say that the utterance in W, 'He is taking the yog for a walk' must mean what we mean by 'He is taking the dog for a walk'. And so it cannot be that the former is true, and the latter false.

Now suppose that we had never been so rash and daring as to make claims of the form

(1) One ought to believe H to degree x

but had instead always settled for the more modest assertion:

(2) According to our inductive practice, one ought to believe H to degree x

If this had been the case there would have been no questioning, as there is with our actual utterances, the truth of those statements. Moreover, we would have been in an instance of W – a world just like ours except for slight linguistic change. One might further imagine that in this W, they abbreviate (2) to the slightly less cumbersome

(3) One ought to believe H to degree x

But, since (2) and (3) are true in W, it follows from the thesis about use and reference that the corresponding claim (1) is true in the actual world. Thus, our inductive practice is correct.

Let me add substance to this idea by responding to a couple of likely objections. In the first place, one might suspect that the argument involves a fallacy of equivocation. For in order that (2) be uncontroversially true in W we must, it seems, assume that its meaning in W is just what we mean by it here. But in order to infer from its truth in W that the corresponding sentence (1) is true here, we have to be in a position to apply the general semantic thesis; so it must be that W differs from the actual world *only* by the designated slight verbal change; thus, it appears that we must assume that the meaning of (2) in W is *not* what we mean by it here, but is rather (as in the dog/yog case) what we mean here by (1). Thus, conflicting assumptions about what (2) means in W are required at different stages of the argument.

However, this objection violates our central presupposition – namely, that once the use of an expression has been completely specified, its reference is then fixed, and no further semantic properties of the expression are left undetermined. In particular, given our characterisation of the physical and mental properties of W, including the way all expressions are used in it, the question of (2)'s meaning does not remain open. Consequently, in order to be sure that the general semantic thesis may be applied to W, it is *not* necessary, antecedently, to assume that (2) in W be given the same significance as (1) in the actual world. We start with the conviction that we would, in fact, find (2) uncontroversial if it were always used instead of (1). This implies that (2) would clearly be true in W. And this, given the general semantic thesis, entails that our assertions of (1) are true.

A second worry might well stem from the observation that philosophers occasionally make such claims as

(4) Even though, according to our inductive practice, one ought not to believe H to degree X, perhaps, nevertheless, one really ought to

Thus W would have to contain, in the same locations, occurrences of

(5) Even though according to our inductive practice one ought not to believe H to degree X, perhaps, nevertheless, really, according to our inductive practice, one ought to

But this would be an absurd claim to make unless 'according to our inductive practice one ought' were ambiguous in W. And in that case, the utterances of (2) are not uncontroversially true.

In this objection I would accept everything but the final step. Just because 'according to our inductive practice one ought', would be ambiguous when uttered by the philosopher in W, it does not follow that in its second occurrence in (5) 'ought' just means what we mean by it. On the contrary, it seems to me that no reason has been given to depart from our original conviction that if we always said (2) rather than (1) then those claims would be un-controversially true. Deviant philosophical usage constitutes notoriously unreliable evidence for semantic theories. What (4) would mean here, and (5) in W, is unclear; but this should not be surprising. Thus, I would regard with suspicion the use of 'one really ought to' in (4) and, similarly, the second occurrence in (5) of 'according to our inductive practice one ought'. They are no more relevant to the normal meaning of (1) than a philosopher's 'He knows but perhaps is not justified in believing it' would be to the concept of knowledge.

I began this chapter by affirming the eminently reasonable idea that our treatment of the problem of induction should include discussions of (A) the nature of our practice, and (B) the conditions of its justifiability. However, if my concluding argument is correct, then the need for such preparation was grossly exaggerated. Nevertheless, some interesting points have emerged from those discussions. We exposed the presuppositions of Carnap's approach: namely, that the constraints on a rational epistemic state are internal, hypothetical, absolute, and pure data constraints. Then we formulated a noncircular demonstrable reliability criterion of justifiability (DR), showed that $c^\dagger$ satisfies it, and that a more audacious practice such as ours must fail to meet it. In addition, I argued that all inductive practices are immodest. In the light of these results a more liberal condition of demonstrable reliability was proposed, which is trivially satisfied by our practice and has no bearing on the question of justification. That was a matter for the final semantic argument.

5

Prediction

If a theory entails that a certain statement is true and it is known that the statement is indeed true, this constitutes a reason to believe the theory. However, various other factors may influence the strength of such evidence. In this chapter I want to explore four superficially related intuitions concerning the degree of confirmation provided by theoretically entailed data: (1) the view that verification of relatively surprising consequences of a theory has especially great evidential value; (2) the idea that survival through severe experimental tests provides a theory with particularly strong support; (3) that the postulation of *ad hoc* hypotheses is generally disreputable; and (4) that the successful prediction of subsequently verified events boosts the credibility of whatever theory is employed to a greater extent than the subsumption or accommodation of previously known results.

It is tempting to regard these intuitions as manifestations of a single underlying phenomenon. For, one might suppose, mere subsumption of known data is inferior to prediction since it is tainted with ad hocness; *ad hoc* hypotheses are undesirable because the facts to which they are tailored have not derived from a genuine test of the new theory; and a genuine, relatively severe test of a theory is one which it is relatively likely to fail – where the data which would constitute passing are relatively improbable and surprising. Thus, our ideas about surprising consequences, severe tests, *ad hoc* hypotheses, and prediction versus subsumption, seem to be intimately related. I will argue, however, that this line of thought is almost wholly mistaken. (3) is explained independently of (1) and (2); and (4) is shown to be incorrect.

Surprise

Let us start with the question: what makes something surprising? Presumably, it is sufficient that we strongly believed, until the moment of truth, that it would not happen. But violation of active expectations cannot be the whole story since we can be reasonably taken aback by things that we had no occurrent beliefs about – explosion of the moon, for example. Should we say then that X would be surprising if either we do believe it won't occur or *would* believe this if we thought about it? Or, better still, X would be surprising if we would assign it a low subjective probability. The trouble with these answers is that they let in too much. Unlikely things are happening constantly, which don't surprise us – things which have as minute a probability as those which do. Suppose I fish a coin from my pocket and begin tossing it. I would be astonished if it landed heads 100 times in a row; but that outcome is no less probable than any other particular sequence of heads and tails; yet certainly not every outcome would surprise me, for example an irregular sequence of about 50 heads and 50 tails. Thus, the improbability of an event is not sufficient – but it does seem necessary. So the problem is to specify what further conditions distinguish improbable events which are, from those which are not, surprising.

To resolve this we should first recognise that our assessment of the subjective improbability of a surprising event derives from our opinions about the circumstances of its occurrence. It is, for example, partly by virtue of the belief that my coin is fair that I assign such a low probability to 100 consecutive heads. Let C represent these beliefs about the circumstances and E be the statement whose truth may or may not be surprising. Then our necessary condition is that our beliefs C are such as to give rise to $P(E) \approx 0$. And, the further condition, which I would like to propose, is that $P(C/E) \ll P(C)$. In other words, the truth of E is surprising only if the supposed circumstances C, which made E seem improbable, are themselves substantially diminished in probability by the truth of E. Now from Bayes' theorem we have

$$P(C/E) = \frac{P(C)P(E/C)}{P(E)}$$

Moreover,

$$P(E) = P(C)P(E/C) + P(K)P(E/K) \\ + P(-C \wedge -K)P(E/-C \wedge -K)$$

where K represents some alternative account of the circumstances which might lead to the truth of E – an account to which we initially assign low credibility. Therefore,

$$P(C/E) \\ = \frac{P(C)P(E/C)}{P(C)P(E/C) + P(K)P(E/K) + P(-C \wedge -K)P(E/-C \wedge -K)}$$

Now, our requirement that $P(C/E) \ll P(C)$ will be satisfied, since $P(C)$ is high, if

$$P(K)P(E/K) \gg P(C)P(E/C)$$

that is, if there is some initially implausible (but not wildly improbable) alternative view K about the circumstances, relative to which E would be highly probable.

To confirm this account let us examine a couple of examples. First, the coin tossing case. Why should 100 consecutive heads be surprising, but not certain other equally improbable sequences? Let C mean 'the coin is fair', E report the surprising outcome, and K designate the unlikely possibility that the coin is heavily biased towards heads or is double headed. Now, for the sake of clarity, let us assume the following not entirely unrealistic subjective probability assignments:

$$0.9 < P(C) < 0.999\,999$$
$$P(E/C) = (1/2)^{100} \approx 10^{-30}$$
$$P(K) > 10^{-10}$$
$$P(E/K) > 10^{-10}$$

It follows that:

$$\frac{P(K)P(E/K)}{P(C)P(E/C)} > 10^{10}$$

and

$$P(C/E) < 10^{-10} \ll P(C)$$

therefore, the truth of E should be surprising.

Compare this with the possible outcome F, a particular irregular sequence of about 50 heads and 50 tails. In this case, there is no obvious candidate for the role of K – no remotely plausible alternative to C, relative to which F would be more probable. We might let K be the hypothesis: the outcome is determined to be F by signals directed from outer space. Then $P(F/K) = 1$, but K is so farfetched and the reference to F, rather than some alternative sequence, so arbitrary that $P(K)$ would be almost negligible – less than $(1/2)^{100}$ – and so there is no reason to expect that $P(C/F)$ is small. Therefore, F would not be surprising.

Second, suppose Jones wins a lottery amongst a billion people (J), and Smith wins three lotteries, each amongst a thousand people (S). Why is Smith's success, unlike Jones', surprising? Well, in each case the favoured background assumption C is that the lotteries are fair, but a possible alternative K is that foul play is involved. The crucial difference between the cases is that $P(S/K) \gg P(J/K)$: on the assumption that the lotteries are fixed, Smith's triple success becomes much more probable than Jones' win. For, in the latter case, the assumption of bias still gives us no reason to expect that Jones, rather than one of the other competitors would be the beneficiary of the bias. So $P(J/K) = P(J/C) = 10^{-9}$. But in the former case, the assumption of bias substantially increases the likelihood that a single individual will win all three lotteries (to, let us say, $1/50$) and so the chances that Smith will be that individual $P(S/K)$ become $\frac{1}{50} \times \frac{1}{1000} = 2 \times 10^{-5}$. Now, if we take the probability that the lotteries are fair, $P(C)$, to be about 0.9, we get:

$$
\begin{aligned}
P(C/J) &= \frac{P(C)P(J/C)}{P(C)P(J/C) + P(K)P(J/K)} \\
&= \frac{0.9 \times 10^{-9}}{0.9 \times 10^{-9} + 0.1 \times 10^{-9}} \\
&= 0.9 \\
&= P(C)
\end{aligned}
$$

and

$$
\begin{aligned}
P(C/S) &= \frac{P(C)P(S/C)}{P(C)P(S/C) + P(K)P(S/K)} \\
&= \frac{0.9 \times 10^{-9}}{0.9 \times 10^{-9} + 0.1 \times 2 \times 10^{-5}} \\
&\approx 0.000\,45 \\
&\ll P(C)
\end{aligned}
$$

Therefore, S, but not J, should be surprising.

Finally, consider the nature of coincidences. They are naturally described as 'just', 'only', or 'mere' coincidences, signifying their misleading import. For a coincidence is an unlikely accidental correspondence between independent facts, which suggests strongly, but in fact falsely, some causal relationship between them. Now, given our proposal, it immediately follows that coincidences are surprising. Thus, it is a surprising coincidence if I accidentally meet a close friend on holiday in Mexico City, for such an encounter suggests a plan; but there is no coincidence when I happen to bump into Mr Samuel Ortcutt from Cleveland, an individual with whom I am not at all acquainted.

These results square well with the proposed account. However, it remains to explain why it is that a theory derives particularly strong support from the accurate prediction of surprising events. The answer is that the evidential value of such predictions stems from the improbability of the predicted events, and is quite independent of that further characteristic, just defended, which renders those events surprising. This is shown by Bayes' theorem, from which we obtain

$$\frac{P(H/E)}{P(H)} = \frac{P(E/H)}{P(E)}$$

When H entails E, $P(E/H) = 1$.

$$\therefore \frac{P(H/E)}{P(H)} = \frac{1}{P(E)}$$

But $P(H/E)/P(H)$ is a measure of the extent to which the discovery of E would raise the probability of H – it is a measure of the extent to which E would provide favourable evidence for H – and this depends solely on $P(E)$. When E is relatively improbable, $1/P(E)$ is relatively large, and the evidence it provides is relatively strong. Now, in order to be surprising, it is not sufficient that E's probability be small. In addition, we must believe that there obtain circumstances C, by virtue of which $P(E)$ is so low, whose probability is substantially diminished by the discovery of E, that is, $P(C/E) \ll P(C)$. However, the satisfaction of this condition would not contribute to E's evidential value.

Severe tests

A severe test of a theory T is an observation designed to discover the truth value of some consequence of T which is antecedently thought to be probably false. In other words, it is a test which T is subjectively likely to fail. Thus, experiments which employ relatively accurate instruments permit more severe tests, since the range of tolerable error is narrower, and so it is less likely that results will conform with theoretical predictions. It is therefore clear, given the previous paragraph, why theories receive particular credit for surviving relatively severe tests.

We can distinguish two kinds of situations in which the prior probability of the predicted data E will be low: (1) when an alternative theory is strongly believed and entails that E is false; (2) when no theory is given much credence, and E is one amongst many possible results of our experiment, for example a particular value of some quantity, such as temperature. Only in the first sort of case would the discovery of E be surprising. Thus, the virtue of surprising predictions is a special case of the virtue of surviving severe tests, which in turn consists simply in the virtue of successfully predicting improbable facts.

Ad hoc hypothesis

An *ad hoc* hypothesis is characterised by Hempel (1966) in the following way:

> [It is introduced] for the sole purpose of saving a hypothesis seriously threatened by adverse evidence; it would not be called for by other findings and, roughly speaking, it leads to no additional test implications.

Example 1.

A simple suction pump which draws water from a well by means of a piston that can be raised in the pump barrel, will lift water no higher than about 34 feet above the surface of the well. Torricelli proposed to explain this by means of the theory that the weight of the atmosphere pushes down on the water surface and is sufficiently heavy to push water up to a maximum of 34 feet. This theory suggested that, since mercury is 14 times as dense as water, the atmosphere should support a mercury column of length 34/14 feet. Pascal recognised a further test implication of this theory: namely, that the length of the column should decrease with increasing altitude, and this again was verified. But now consider an alternative theory: that nature abhors a vacuum. This also explains the working of suction pumps, and so on. But in contrast to Torricelli's theory, it gives no reason to suppose that the length of a mercury column would decrease with increasing altitude. Such a decrease is observed; but does this falsify the hypothesis that 'nature abhors a vacuum'? No, for the threatening data may be accommodated by the *ad hoc* hypothesis that nature's hatred of vacuums decreases with increasing altitudes.

Example 2.

There was an eighteenth-century view according to which the combustion of metals involved the escape of a substance called phlogiston. This theory was threatened by Lavoisier's experiments which showed that the products of combustion were not lighter than the original metal – as would have been expected on the phlogiston theory – but were heavier. However, some proponents of the phlogiston theory attempted to save their theory in the face of Lavoisier's results, by postulating the *ad hoc* hypothesis: phlogiston has negative weight.

Now let us raise the following questions:

(a) Do the examples fit Hempel's general characterisation of an *ad hoc* hypothesis?
(b) Why would an *ad hoc* hypothesis *in Hempel's sense* be objectionable: something whose postulation a rational investigator should avoid?
(c) Is it the case and, if so, why is it that the hypotheses in Hempel's examples are objectionable?
(d) Can we formulate a better definition of '*ad hoc* hypothesis' and explain why such hypotheses should be avoided?

Hempel maintains that an *ad hoc* hypothesis is introduced merely to save a theory and is not called for by other findings. This much seems clearly true of his examples. However, he says, in addition, that an *ad hoc* hypothesis leads to no additional test implications; and it is far from clear that this applies to the paradigm cases. From the claim that nature's abhorrence of a vacuum decreases with altitude, we infer that whenever place P_1 is higher than P_2, the length of a mercury column at P_1 will be less than one at P_2. Also, one might expect other cases of altitude-related variation in nature's powers. From the claim that phlogiston has negative weight, we might conclude that if a metal is burned in a closed container, and some method designed to remove *only* the material products of combustion, then the vessel will weigh less than an empty vessel (since it will contain a quantity of phlogiston). Thus, it seems that Hempel's examples don't quite conform to his general characterisation of ad hocness.

Now let us consider what *would* be wrong with postulating a hypothesis which *does* fit his general characterisation of ad hocness. It can hardly be regarded as illegitimate in general to save one's hypothesis by blaming a mistaken prediction on one of the auxiliary hypotheses. For something must be given up, and if the main hypothesis were abandoned we might just as easily be accused of ad hocness in our decision to save the auxiliary hypotheses in the face of evidence which threatens to falsify them. So, if there is anything wrong with the postulation of *ad hoc* hypotheses in Hempel's sense, it must be because such hypotheses lead to no additional test implications. Now this would indeed be bad; for without further test implications we can accumulate no reasons to believe the hypothesis is true. So we can concede that *ad hoc* hypotheses, in Hempel's sense should be avoided.

But we are now left with a dilemma. For, on the one hand, it is natural to agree that there is something undesirably *ad hoc* about the hypotheses in

Hempel's examples: that nature abhors a vacuum, and that phlogiston has negative weight. But, on the other hand, these cases fail to fit precisely that clause of Hempel's general definition of ad hocness, which would make the postulation of *ad hoc* hypotheses objectionable. This suggests that Hempel has not accurately located what it is about *ad hoc* hypotheses that makes them objectionable. But if the objectionable feature is not, as Hempel says, that such hypotheses have no further test implications, what then is it?

One notable feature of Hempel's examples is that in each case there was an alternative to the *ad hoc* manoeuvre: that nature abhors vacuums less at greater altitudes is proposed in resistance to Torricelli's theory of atmospheric pressure; that phlogiston has negative weight is postulated to save the phlogiston theory from being replaced by the oxygen account of combustion. This suggests that what marks a hypothesis as objectionably *ad hoc* is the availability of some alternative theory which could equally accommodate the data and which is, in addition, in some way more desirable than the modified old theory. According to this view, the ad hocness of a hypothesis depends primarily upon the intrinsic plausibility, simplicity, or prior probability of the overall theory in which it is included. When theory T is embarrassed by adverse data which may be accommodated by a revision to T', and there is an alternative theory R which accounts for the same data as T', then acceptance of the modification is regarded as *ad hoc* just in case T' is intrinsically less plausible than R. Thus, the existence or absence of further test implications is irrelevant. It is quite possible for T' to have observable consequences which are distinct from those of T, and distinct from those which provoked the revision.

This account may seem to require that we cannot determine whether the postulation of a hypothesis is *ad hoc* merely by examination of the hypothesis in question, the theory it is designed to save, and the data; but would need to know something about available alternatives. I want to resist this plausible line of thought. Acceptance of a theory may be *ad hoc* even in the absence of decent alternatives. It is sufficient that the theory have intrinsic implausibility. The presence of more desirable alternatives will serve to highlight the ad hocness involved in accepting our rigged up account; it will enable us to recognise it more easily; but it is not essential. The awkward contortions of contemporary 'flat earth theory', required to accommodate such phenomena as apparent circumnavigation, are sufficiently extreme to render it ludicrously *ad hoc*. Its intrinsic undesirability is particularly glaring in contrast with our alternative, but in itself it constitutes sufficient grounds for scepticism.

Literally, something is *ad hoc* just in case it is specifically designed for some particular purpose. When applied to hypotheses, the relevant purpose is the accommodation of evidence. Thus, an element of a theory is said to be *ad hoc* when it is included solely in order that the theory will entail certain statements. This implies that there is no ulterior reason for the inclusion of that element. It is not a natural component, does not harmonise with the other elements and produces a whole which is contrived, complex, and assigned a low credibility in our inductive practice. In other words, a theory is *ad hoc* if and only if its formulation is motivated by nothing more than a desire to accommodate certain facts. Such a theory cannot have a high prior probability or else that feature would constitute a further reason for its postulation. Therefore, *ad hoc* theories are implausible.

Prediction versus accommodation

Evidently, there is a certain symmetry between prediction and accommodation: crudely, prediction is the deduction of some future condition, whereas accommodation is the deduction of some past or present conditions. Perhaps it would be better to formulate the distinction in terms of unknown and known, rather than future and past conditions; for we can say that a theory makes predictions about the origins of the universe. The symmetry would then be characterised as follows: that both prediction and accommodation consist in the deduction of some fact; and that we apply the term 'prediction' when the fact has not been established, and the term 'accommodation' when it has.

I wish to consider whether more reason to believe a theory is provided by its correct predictions than by its accommodation of data. It is clear that usually neither would result in conclusive confirmation. For our theory may not be unique in accurately predicting an event nor in being able to accommodate its occurrence. Nevertheless, both prediction and accommodation have some evidential value, and the question is, which is better? Also, it is plain that some predictions are more impressive than some accommodations, and vice versa. So our problem might be formulated more precisely as follows.

Compare two possible circumstances. In the first, which I shall call P (for prediction), we have some reason, provided by evidence E, to believe a theory T, and we use T to predict that a certain experiment will yield result D. That is, we deduce D from T, then do the experiment, and discover to our delight that D is indeed the case.

In the second circumstance A (for accommodation), we again have the information represented by E, but formulate no theory to account for it; we feel that more evidence is needed. So we do the same experiment and discover D. At this point we scout around for a theory which will accommodate our total data $E \wedge D$, and theory T is the one we arrive at.

In the first case, T is formulated to account for data E and used to predict D which is then experimentally verified. In the second case, the theory T is formulated only after, and in order to account for, the discovery of $E \wedge D$. The question: do investigators in P have, at the end of the day, more reason to believe the theory than investigators in A?

For example, suppose that we want to discover the relationship between two experimentally determinable quantities X and Y. Let E be a set of five data points and suppose that a straight line can be drawn through them. Let T be this straight line hypothesis.

Now let D be a set of five more data points which also lie on the straight line T. Would T be better confirmed if it were formulated on the basis of the first five points and then used to predict for each value of X, in the set of D, the correct value of Y, than if it were formulated after all the ten points were found, and specifically tailored to fit them?

I think it is widely believed that the answer, both in this example and generally, is yes: the verification of predictions is a vital mode of theory confirmation since it carries substantially more evidential weight than mere accommodation of data. I shall argue that this is wrong. First, I want to show briefly that it is unrelated to, and derives no support from, our intuitions about surprising evidence, severe tests, and *ad hoc* postulates. And second, I will examine in detail what I think is the real source of the view.

As we have seen, the verification of relatively surprising predictions and the passing of severe tests obtain their confirmation power from a combination of (a) the fact that the resulting data are antecedently improbable, and (b) the consequence of Bayes' theorem:

$$\text{If } H \text{ entails } E_1 \text{ and } E_2, \text{ then} \frac{P(H/E_1)}{P(H/E_2)} = \frac{P(E_2)}{P(E_1)}$$

But this account gives no reasons to attach particular value to prediction. It is true that in the predictive case our degree of belief in H should increase, with the discovery of D, from $P_E(H)$ to $P_E(H/D)$ – where $P_E(q)$ represents our beliefs before the experiment. But it is perfectly compatible with this to suppose that the appropriate degree of belief in H, given its ability to accommodate $E \wedge D$, is also equal to $P_E(H/D)$. For, suppose there is a function $c(H/q)$ which specifies the degree to which H is confirmed by variable evidence, q. Then, $P_E(H)$ should be $c(H/E)$; and $P_E(H/D)$ should be $c(H/E \wedge D)$ which should also equal the ultimate degree of belief in H in the case of accommodation.

Nor may the intuition in question be derived from the undesirability of *ad hoc* postulates. For, according to the above discussion, this undesirability is engendered by the intrinsic prior improbability of such theories, due to features such as complexity which are independent of their compatibility with data. But the question before us concerns the relative evidential value of a single set of data $E \wedge D$, with respect to a single theory T, in two possible circumstances, prediction and accommodation. Thus, insofar as T is undesirably *ad hoc* in the case of accommodation, its intrinsic prior improbability must equally preclude the achievement of substantial credibility in the predictive case. In short, the ad hocness of a theory may only account for its low credibility compared with other theories. It is irrelevant to questions about the variable support provided by different factors for a single theory.

Let me now turn to the argument which does, I think, underlie the view that prediction has more evidential value than accommodation. Suppose we ask ourselves, in each of the two situations P and A, why does our theory fit the facts D? In the second case A, the best explanation for this coincidence is that our theory was specifically tailored to fit them. In the first case P, the best explanation is that the theory fits the facts (that is, enables an accurate prediction of them) because it is true. Thus, via inference to the best explanation, we may infer, in P, that T is true; but no such inference is sanctioned in A, for here the truth of T is not as good an explanation of the fit as the fact that our theory was deliberately designed to fit the data – we wouldn't have selected a theory which didn't fit.

Putting the point again, we might say that when a theory accurately predicts the results of an experiment, we are provided with some reason to believe it because we need an explanation of its success and the best explanation of its success is that it is true. But if the theory is merely rigged up to fit the data, which has all been obtained in advance – if it is specifically formulated so as to accommodate that data – then we know precisely why it fits the data, and there is no longer required any further explanation of this fact.

Analogously, suppose Smith's car won't start and he wonders why. He has ruled out all but two possible explanations:

H_G : he is out of gas

H_S : the starter is broken

and believes these to be equally likely. Now, suppose he discovers that indeed the tank is empty. This, of course, does not establish that the starter is healthy. The breakdown may be overdetermined. But clearly, it would not be reasonable for him to attach the same probability to the hypothesis that the starter is broken *after* he finds he is out of gas, as he did before. It becomes much less likely. This is because the initially substantial probability which attached to the broken starter hypothesis derived from the possibility that it might be needed to account for the car's failure to start. Once it becomes clear that another hypothesis, which would account perfectly for this, if it were true, is in fact true, the probability that the starter is broken diminishes to the value it would have had before Smith discovered that his car wouldn't start.

Similarly, the hypothesis that our theory was designed to fit $E \wedge D$, would perfectly well account for the fit, if it were known to be true; and in case A, it is in fact known to be true. So the probability that the theory T is correct, which would also account for the fit if T were known to be true, reduces to the probability we assigned it before the data were obtained. Thus, in case A, our theory receives no support from its accommodation of D, whereas in case P, the accurate prediction of D provides substantial confirmation.

To see what is wrong with this argument, consider the following example: why is it that my car is green? Is it because I always insist on green cars? Or is it because the previous owner painted it green? Obviously, in this case, a demonstration of the truth of one of those hypotheses would not diminish the probability that the other is also correct. Unlike the broken starter and empty tank hypotheses, which in some sense compete with one another to explain the car's failure to start, the hypotheses under present consideration do not strike us in the same way as alternatives. Rather, it seems that neither hypothesis can provide us with as full an explanation of why it is that my car is green as can their conjunction.

Now in the situation A, where our theory has been tailored to the facts, we compared two possible explanations of this fit: (1) that our theory is true, and (2) that our theory was formulated to fit the facts. According to the argument above, account (2) is the better of these explanations, and it is concluded that the fit, having been explained by (2), provides no reason to believe (1) that our theory T is true. My view is that this is incorrect, and that the explanation provided by (2) neither precludes the truth of (1), nor renders it less probable or valuable as an explanation. I think that the relevant analogy is provided by the accounts of why my car is green, and not by the two hypotheses formulated to explain why Smith's car doesn't start.

These claims are illuminated and justifiable within a Bayesian framework. First, again consider Smith's car. The empty tank and broken starter hypotheses offer alternative descriptions of the cause of its failure to start. And we can explain as follows the fact that verification of one hypothesis should diminish the probability of the other:

Let

$$H_G = \text{there is no gas}$$
$$H_S = \text{the starter is broken}$$
$$F = \text{the car fails to start}$$

From Bayes' theorem:

$$P(H_G/F) = \frac{P(H_G)P(F/H_G)}{P(F)}$$

where the subjective probabilities are those assigned by Smith before he attempts to start his car. Now

$$P(F) \ll 1$$

and

$$P(F/H_G) = 1$$
$$\therefore \ P(H_G/F) \gg P(H_G)$$

that is, the probability of no gas is substantially increased with the discovery that the car won't start. Furthermore,

$$P(H_G/F \wedge H_S) = \frac{P(F/H_G \wedge H_S)P(H_G/H_S)}{P(F/H_S)}$$

and

$$P(F/H_S \wedge H_G) = 1$$

and

$$P(F/H_S) = 1$$
$$\therefore \ P(H_G/F \wedge H_S) = P(H_G/H_S)$$

But the probability of having no gas is not affected by the supposition that the starter is broken. That is,

$$P(H_G/H_S) = P(H_G)$$
$$\therefore\ P(H_G/F \wedge H_S) = P(H_G)$$

In other words, the further discovery that the starter is broken should neutralise the original effect of the failure to start.

Now, for the second case, let

H_P = the previous owner painted the car an ineradicable green

H_I = I insisted upon a green car

F = my car is green

As before, we have

$$P(H_P/F) = \frac{P(H_P)P(F/H_P)}{P(F)}$$

where the subjective probabilities are those assigned by someone unaware of the truth of F. Thus,

$$P(F) < 1$$
$$P(F/H_P) = 1$$
$$\therefore\ P(H_P/F) \gg P(H_P)$$

Now, let us consider, paralleling the previous analysis, the effect of the further discovery of H_I:

$$P(H_P/H_I \wedge F) = \frac{P(H_P/H_I)P(F/H_P \wedge H_I)}{P(F/H_I)}$$

As before,

$$P(F/H_P \wedge H_I) = 1$$

and

$$P(F/H_I) = 1$$
$$\therefore\ P(H_P/H_I \wedge F) = P(H_P/H_I)$$

But we cannot repeat our earlier move and identify $P(H_P/H_I)$ with $P(H_P)$. The probability of a broken starter is not affected by data to the effect that there is no gas. But the probability that the previous owner of my car painted it green *is* increased by data to the effect that I insisted on having a green car. For that data informs us that my car is green, and improves the chances of all accounts of how it might have come to be that way. Moreover, the only element in this data which is relevant to the probability of H_P is that my car is green.

$$\therefore\ P(H_P/H_I \wedge F) = P(H_P/F)$$

Finally, let us examine the situation A. Our concern is to explain the fact

$$F : \text{our theory fits the data } D$$

and we have under consideration the hypotheses

$H_T = $ our theory is true
$H_R = $ we required a theory which would fit the data D

As in the previous examples we can show that

$$P(H_T/F) \gg P(H_T)$$

that is, if all we know is that our theory fits the data, this provides us with further reason to believe that it is true. The critical question is whether or not the additional knowledge which we have in the context A of accommodation – namely, that we required a theory which would fit the data D – undermines this evidence. Thus, we are concerned with the value of $P(H_T/F \wedge H_R)$. Now, just as in the previous treatments, we have from Bayes' theorem:

$$P(H_T/F \wedge H_R) = \frac{P(H_T/H_R)P(F/H_T \wedge H_R)}{P(F/H_R)}$$

Also,

$$P(F/H_T \wedge H_R) = 1$$

and

$$P(F/H_R) = 1$$
$$\therefore\ P(H_T/F \wedge H_R) = P(H_T/H_R)$$

The proponent of the argument which we are attempting to rebut maintains that the knowledge of H_R undermines the reason to believe H_T, which would otherwise be provided by F. This would require that $P(H_T/H_R)$ be identified with $P(H_T)$ – a move which we have seen is legitimate in the case of the broken starter and no gas hypotheses, but which is mistaken in the second case where the hypotheses are probabilistically dependent upon one another.

Now, it is clear that the hypotheses H_T and H_R fall into the second group. If we know of our theory that it was required to fit data D, this *does* provide us with more reason than we would otherwise have had to believe that this theory is true. Thus, $P(H_T/H_R) \neq P(H_T)$. As in the previous example, what is relevant to the probability of H_T, in the information H_R, is precisely that our theory fits the data D. Therefore, $P(H_T/H_R) = P(H_T/F)$, and so

$$P(H_T/F \wedge H_R) = P(H_T/F)$$

Thus, the capacity of our theory to accommodate the data D provides just as much reason to believe it as its ability in context P to predict those data.

I conclude this chapter with one more argument for this view. Suppose we know that our theory T entails the established data $E \wedge D$, but we don't know whether they are predicted or accommodated. Consider the different effects of the following two possible discoveries upon the probability of T:

(B): T was formulated to account for E, further consequences D were deduced from T, an experiment was done to determine the truth value of D, and D was found to be true

(A): T was formulated to account for $E \wedge D$

We have

$$P(B) + P(A) = 1$$

and

$$P(B/T) + P(A/T) = 1$$

and from Bayes' theorem

$$P(T/B) = \frac{P(T)P(B/T)}{P(B)}$$

Therefore, any special virtue of prediction depends upon the relationship between $P(B)$ and $P(B/T)$. Does it raise the likelihood of B (and so diminish that of A) to suppose that our theory T is true?

Assuming that we have a propensity to pick true theories – that the truth looks plausible to us – the answer may appear to be yes. For to suppose that T is true is to suppose that probably T looks plausible; and this tends to favour the possibility that T would have recommended itself to our attention, even before the discovery of D: thus $P(B/T) > P(B)$. But if so, we infer from Bayes' theorem:

$$P(T/B) > P(T)$$

and consequently

$$P(T/A) < P(T)$$

And so prediction does appear to have particular evidential value.

Note, however, the implicit assumption that we are in the position of *inferring* T's inherent plausibility. Not knowing exactly what the theory is, our confidence in its apparent plausibility is increased by supposing it is true. But if we do actually have the theory before us and, in full possession of our rational faculties, can simply *see* how plausible it looks, then the preceding argument is fallacious: it is not then the case that supposing the *truth* of T will modify our estimate of its plausibility. Thus,

$$P(B/T) = P(B)$$

and therefore

$$P(T/B) = P(T)$$

It is of no consequence that the data were predicted.

Thus, information to the effect that a theory correctly predicted certain facts may enhance its credibility – in some circles – by more than the news of its successful accommodation of the same data; but only amongst those people who are not acquainted with the theory – who know it, perhaps, as 'T' or 'Freud's theory', yet are unfamiliar with its principles. But if, on the other hand, we grasp the theory and can rationally assess its plausibility, then no information about whether the entailed data were predicted or accommodated should have the slightest evidential value.

6

Evidence

The evidential value of varied data

It is an undeniable element of scientific methodology that theories are better confirmed by a broad variety of different sorts of evidence than by a narrow and repetitive set of data. To support a quantitative law, we want numerical data which include both high and low values of the magnitudes which it relates; and, similarly, a hypothesis such as 'All ravens are black' is best confirmed if the observed ravens have been selected from a wide range of circumstances. I would like to offer a Bayesian justification for these truisms – to provide a relatively clear explication of what it is for the evidence for a hypothesis to be diverse, and to demonstrate that such evidence increases the probability of the hypothesis by more than a uniform data set.

The intuitive idea which underlies what follows is: diverse data tend to eliminate from consideration many of the initially most plausible, competing hypotheses. Narrow data, on the other hand, leave many initially very plausible alternatives in the field, and can therefore provide relatively little reason to select any one of them. More precisely, I want to suggest that evidence is significantly diverse to the extent that its likelihood is low, relative to many of the most plausible competing hypotheses.

If this is right, we can easily explain the power of diverse evidence. Start with Bayes' theorem:

$$P(H_1/E) = \frac{P(H_1)P(E/H_1)}{P(H_1)P(E/H_1) + P(H_2)P(E/H_2) + \cdots + P(H_k)P(E/H_k)}$$

where $H_1, H_2, \ldots, H_k$ are mutually exclusive and it is known that one of them is true. Next, suppose that E_D and E_N are both entailed by H_1, but that E_D contains a wide spectrum of information whereas E_N is a narrow set of data. Bayes' theorem yields:

$$\frac{P(H_1/E_D)}{P(H_1/E_N)} = \frac{P(H_1) + P(H_2)P(E_N/H_2) + \cdots + P(H_k)P(E_N/H_k)}{P(H_1) + P(H_2)P(E_D/H_2) + \cdots + P(H_k)P(E_D/H_k)}$$

Now, given my account of diversity, the probability of E_D is low, relative to many of those hypotheses with a high prior probability. In other words, there are many cases of H_j such that $P(H_j)$ is substantial and $P(E_D/H_j)$ is less than $P(E_N/H_j)$. Therefore we expect that

$$P(H_1/E_D) > P(H_1/E_N)$$

The diverse data set provides more support than the narrow one.

A good example of this phenomenon is provided by curve fitting (see Fig. 10). Consider the data sets E_D and E_N, where E_D contains three points spread far apart, which could, within the limits of experimental error, be connected by a straight line, and E_N also contains three points connectable by the very same line but bunched together. Let the straight-line hypothesis be H_s. The question is: why do we feel that E_D provides better support for H_s than does E_N? And our answer is that E_D tends to exclude more of the more plausible competitors.

We associate plausibility with simplicity. We think that straight-line hypotheses are more likely to be true than hypotheses compatible with the same data points, expressed by high-order polynomials, trigonometric functions and yet more complex functions. After straight lines, slow-breaking hyperbolae rank next in plausibility. Now, both E_D and E_N may be connected by various gentle curves – slow-breaking hyperbolae. Since the points in E_N are relatively bunched together, they are compatible, given experimental error, with more such curves than the points in E_D. In fact, any such curve which could accommodate E_D could also accommodate E_N; but there are many slow-breaking hyperbolae which fit E_N but which curve too dramatically to pass near the extreme points of

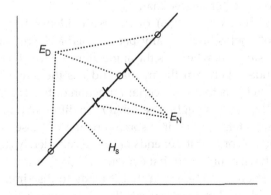

Fig. 10

E_D. Thus, E_D excludes more of the most plausible alternatives to H_s and E_N. Consequently, the probability of H_s is increased more substantially by E_D than by E_N.

To show this formally, employ Bayes' theorem to obtain:

$$\frac{P(H_s/E_D)}{P(H_s/E_N)} = \frac{P(H_s) + P(H_b i)P(E_N/H_b i) + \cdots + P(H_c i)P(E_N/H_c i) + \cdots}{P(H_s) + P(H_b i)P(E_D/H_b i) + \cdots + P(H_c i)P(E_D/H_c i) + \cdots}$$

where $H_b i$ are the straight lines (other than H_s), the slow-breaking hyperbolae and the other gentle curves to which we assign non-zero credibility; and $H_c i$ are the high-order polynomials and other crazy curves. Now, the prior probabilities of H_s and the H_b are so great compared to those of the H_c that we may neglect the terms involving $P(H_c)$. There are a handful of crazy hypotheses $(H_c i)$ which go near E_N and equally few equally improbable hypotheses which go near E_D. But the prior probability that any of these (that is, their disjunction) is true, is negligible. Thus the sums in both numerator and denominator of those terms involving $P(H_c i)$ will be small and roughly equal.

Furthermore, for any $H_b i$

$$P(E_D/H_b i) \leq P(E_N/H_b i)$$

since any straight line or slow curve near the points of E_D will pass near the points of E_N. And, for many $H_b i$,

$$P(E_D/H_b i) \ll P(E_N/H_b i)$$

since there are straight lines and slow curves through E_N which go nowhere near E_D. Therefore, the denominator of the right-hand side of the above equation must be smaller than the numerator, and so E_D confirms H_s to a greater degree than E_N.

In Hempel's discussion (1966) of diverse evidence he considers the confirmation of Snell's law: for any pair of media, there is a constant μ such that $\sin i/\sin r = \mu$, where i is the angle of incidence of a beam of light upon the boundary between the media, and r is the angle of refraction. This hypothesis claims that r is a certain function of two parameters i and μ. Alternative hypotheses might assert that r is a different function of those very same parameters, or that it is some function of those, together with further parameters, or that it depends on some completely different set of magnitudes. I have maintained that evidence for Snell's law is diverse and, consequently, strong, to the extent that it tends to discriminate amongst the most plausible of these alternatives. This is borne out of consideration of the following data sets:

(D_1) i is fixed at 30°, r is measured; our experiment is repeated 100 times on a single pair of media.

(D_2) same as above, except that 50 of the experiments are done with $i = 20°$.

(D_3) same as D_2, except that the temperature of the media is varied between experiments.

(D_4) same as D_2, except that the experimenter's eye colour is varied.

Now, there is a technical sense in which each of these data sets is equally diverse: each is produced by 100 different experiments in different conditions. Nevertheless, we would naturally regard D_2 as more varied than D_1, and D_3 (but not D_4) as more varied than D_2. This is because we identify variation with significant variation. We don't suppose that the variation in the time at which an experiment is conducted is significant. Nor do we think that the experimenter's eye colour matters. Another way of putting this is to say that we assign a low prior probability to hypotheses according to which r depends upon time or upon experimenter's eye colour. Thus, data sets which preclude many such hypotheses are not regarded as diverse and do not possess great evidential value. For this reason, D_4 provides no more support than D_2. On the other hand, D_3 goes significantly beyond D_2. For it is not initially implausible that r should depend upon the temperature of the media. Thus, the results of D_3 exclude a class of hypotheses with a significant prior probability and, therefore, confer further credibility upon Snell's law.

In summary, I have argued that evidence for a hypothesis is regarded as significantly diverse to the extent that it is improbable, relative to those alternative hypotheses with high prior probabilities; and I have explained why such evidence is stronger. Thus, I deny that a data set can be evaluated with respect to significant diversity unless this is done in relation to a particular class of alternative hypotheses and prior assessment of the plausibility of those hypotheses. More modestly, I should say that no such absolute measure of diversity will imply the usual methodological characteristic of varied evidence: namely, greater evidential value. Alternatively, we could conclude that diversity, in itself, is of no interest; and that it is, strictly speaking, a mistake to suppose that varied evidence is better. Only evidence which is *significantly* diverse, in the sense characterised above, has any special evidential value.

The value of further data

Suppose we are concerned about the truth of a certain hypothesis H, and possess various items of information $E_1, E_2,...,E_k$ which bear upon the question. We accept the Carnapian confirmation function c and know that $c(H/E_1) = x_1$, $c(H/E_1 \wedge E_2) = x_2$, ... and $c(H/E_1 \wedge \cdots \wedge E_k) = x_k$. What should be our degree of belief in H?

The obvious answer appears to be x_k. Our degrees of belief should be based upon *all* the evidence at hand. However, some philosophers, following A. J. Ayer (1972), have thought there to be a difficulty in providing justification for this intuition. This matter is known as the problem of total evidence: why should our expectations be based upon all of our data rather than just some part of it?

Related to this is the so-called problem of further evidence. We not only feel that all the available evidence should be employed, but also that additional information should be acquired. Of course, it may in certain cases be extremely inconvenient to gather some items of data which we would like to have, and we may conclude that it isn't worth the trouble to do so. But this does not detract from the inclination to suppose that the extra facts would be of epistemological value.

The problem of further evidence specifically concerns the justification for our intuition that *further* data should be acquired. But it is natural to suspect that the two problems have a single solution: namely, some explanation of why it is better to base expectations upon as much evidence as possible. With such an account in hand, it seems that we could easily justify (a) the use of all, rather than merely some, of our present data, and (b) the inclination to gather more data.

However, I think that this temptation to merge the questions should be resisted. For the problem of total evidence is nonexistent.

If the logical probability of H, given E, is x, this entails, by definition of 'logical probability', that, in the evidential circumstances where just E is known, one's degree of belief in H ought to be x. The notion of logical probability is specifically designed to have this feature: that degrees of logical probabilification by E correspond to the degrees of belief that are prescribed by reason for the state of total evidence E. So, if $c(q/p) = x$ is the correct confirmation function, it follows immediately from the very meaning of that claim that if $E_1 \wedge E_2 \wedge \ldots \wedge E_k$ constitutes our evidence, then $c(H/E_1 \wedge E_2 \wedge \ldots \wedge E_k)$ should be our degree of belief in H. One might of course wonder, quite legitimately, if H's degree of confirmation relative to $E_1 \wedge E_2 \wedge \ldots \wedge E_k$ is really x_k and not, perhaps x_1 instead, or x_2; one might challenge the intuitive relevance of our last bits of data and advocate conformity to a different confirmation function c' which is such that $c'(H/E_1 \wedge E_2 \wedge \cdots \wedge E_k) = c'(H/E_1) = x_1$ .But these questions belong squarely within the traditional problem of induction. They are questions about the merits of our inductive practice compared to others and, so far as I can see, there is no additional problem of total evidence lurking in the background.

However, the problem of further evidence cannot be dismissed so lightly. Suppose my degrees of belief are

$$P(H) = x \quad 0 < x < 1$$
$$P(H/E) = y \quad 0 < y < 1$$
$$P(H/-E) = z \quad z \neq y$$

If these, given my evidential circumstances, are prescribed by the correct confirmation function, then those degrees of belief are as they should be. But suppose I am in a position to discover the truth value of E. I realise that if I knew E's truth value, then my degree of belief in H would have to be other than it is. But this does not immediately entail that it is reasonable for me to determine E's truth value. It only entails that I would have to change my degree of belief in H *if* I conducted that investigation.

Thus, the problems of total and further evidence are not simple variants of a single underlying problem: to rationalise our fondness for evidence. The total evidence question is trivial. The desire to gather extra information remains to be justified. Let us consider some possibilities.

(1) It is worth noting that the superior confirmation power of *diverse* data, explained above, is a separate matter and of no help to us here. I have shown that if H entails both E_1 and E_2, and if E_1 is more diverse than E_2, then E_1 will confirm H to a greater degree than E_2. But it by no means follows that further information is worth acquiring. Suppose we are concerned with some hypothesis H, which entails E. To discover the *truth* of E may further diversify the evidence for H. But since we don't know in advance whether E will turn out to be true, we cannot assume that the new total evidence will determine an increase in the credibility of H.

(2) It might be thought that the value of new information may be derived from (a) our desire to know definitely whether hypotheses are true or false, and (b) the fact that, as more evidence accumulates, the probabilities of hypotheses tend towards 1 or 0. However, this explanation will not do. Even if claim (b) could be formulated with precision, and substantiated, it certainly cannot be supposed that the discovery of E's truth must move the probability of H towards its ultimate value. It is quite possible that H is in fact false, even though $P(H/E) > P(H)$. For example, let us suppose that although there is no life on Mars, a soil extract reveals all essential ingredients. Thus $P(H/E)$ is misleadingly greater than $P(H)$. Now, if (b) is roughly correct then accumulating data would eventually point us in the right direction and reduce $P(H)$ to near zero. However, we have no indication as to when the big fluctuations will cease and the slide to zero occur. We don't know whether our next observation will move $P(H)$ in the right or the wrong direction, so why should we bother to make it?

Nor is it the case, in general, that the probability of H is *probably* moved towards its ultimate value. To see this, simply add $P(H/E) < 1/2$ to the previous suppositions. Then the probability that the probability of H has been moved towards its ultimate value is equal to the probability that 1 is that ultimate value and this is less than $1/2$. If the probability of life on Mars, given our soil sample evidence, goes up from $1/10$ to $1/3$, then we still have to suppose that there is probably no life on Mars and that the probability of life on Mars has probably been moved in the wrong direction by the evidence. Despite these difficulties, I think this approach is on the right track, and I will return to it later.

(3)   It may seem that when a range of alternative hypotheses $H_1, H_2, \ldots,$ H_k are in question, then, as more evidence accumulates, the most probable of these hypotheses – whose identity varies – becomes increasingly likely to be true. In other words, suppose H_1 is the most credible right now; H_2 the most probable, given E; and H_3 the most probable, given $-E$. Then, we might think that both $P(H_2/E)$ and $P(H_3/-E)$ must be greater than $P(H_1)$. So that the discovery of $E's$ truth value, whatever it turns out to be, will justify some greater degree of conviction. This rationale is particularly plausible in the light of examples in which characteristics of a population are inferred from the examination of random samples. For, suppose we want to know the proportion of things from a population with some feature. The best guess is that this equals the proportion of things with the feature in our random sample. Moreover, it can be shown that, although what qualifies as the most probable hypothesis may change as the sample is augmented, the probability of whatever is the most probable hypothesis must steadily increase. In such cases, therefore, we do have a clear motive to acquire new data. But not in general. For there is no reason to rule out the possibility of cases in which a discovery will not have this effect. It may be, for example, that our range of alternative hypotheses consists of just H and $-H$; and that

$$P(-H/-E) < P(-H) < P(-H/E) < P(H/E) < P(H) < P(H/-E)$$

and here the probability of the most probable hypotheses will increase only if E turns out to be false.

(4)   There are circumstances in which expected utility is promoted by the acquisition of data. This approach has been explored by Keynes (1921), Good (1966), Rosenkrantz (1977) and others. Suppose we must choose between actions X and Y; and that the expected value of X is greater than the expected value of Y, so that X is preferable. I want to compare the expected utility of simply performing this preferred act, to the expected utility of first finding out the truth value of E and then doing whichever act is preferable in the light of the new information.

Let $S_1, S_2, \ldots, S_i, \ldots, S_n$ be a set of mutually exclusive and exhaustive, possible future states of affairs (possible consequences of X and Y); and let $V(S_i)$ represent our gain if S_i will be the actual state. Then, the expected value of X

$$V(X) = \Sigma P(S_i/X)V(S_i)$$
$$V(Y) = \Sigma P(S_i/Y)V(S_i)$$

and, by hypothesis, $V(X) \geq V(Y)$

$$\therefore \sum [P(S_i/X) - P(S_i/Y)]V(S_i) > 0$$

Suppose that the discovery of E would so affect the probabilities that Y's expected utility would become greater than X's, that is,

$$V_E(Y) > V_E(X)$$

that is,

$$\Sigma P(S_i/YE)V(S_i) > \Sigma P(S_i/XE)V(S_i)$$

It follows, as we shall now see, that the discovery that E is false would leave unchanged the ranking of expected utilities. For we have from probability theory

$$P(S_i/X) = P(E/X)P(S_i/XE) + P(-E/X)P(S_i/X(-E))$$

Also, the present truth or falsity of E is probabilistically independent of whether X or Y is performed.

$$P(E/X) = P(E/Y) = P(E)$$
$$P(-E/X) = P(-E/Y) = P(-E)$$
$$\therefore V(X) = \Sigma [P(E)P(S_i/XE) + P(-E)P(S_i/X)(-E)]V(S_i)$$

and

$$V(Y) = \Sigma [P(E)P(S_i/YE) + P(-E)P(S_i/Y(-E))]V(S_i)$$

But

$$V(X) > V(Y)$$
$$\therefore P(E)\Sigma[P(S_i/XE) - P(S_i/YE)]V(S_i)$$
$$> P(-E)\Sigma[P(S_i/Y(-E)) - P(S_i/X(-E))]V(S_i)$$

But $V_E(Y) > V_E(X)$ and, therefore, the left-hand side is negative: consequently the right-hand side is negative.

$$\therefore V_{-E}(Y) < V_{-E}(X)$$

Thus, in our present state of knowledge, X is the best thing to do, and it would still be best if $-E$ were discovered; but if E were known then Y would be better than X.

Now, let us compare the expected values of

(α) simply perform whichever is better of X and Y;
(β) first determine the truth value of E, and then perform whichever is better in the light of the new information

$$V(\alpha) = \text{the value of whichever is better of } X \text{ and } Y$$
$$= V(X)$$
$$= \Sigma P(S_i/X)V(S_i)$$
$$= \Sigma[P(E)P(S_i/XE) + P(-E)P(S_i/X(-E))]V(S_i)$$
$$= P(E)\Sigma P(S_i/XE)V(S_i) + P(-E)\Sigma P(S_i/X(-E))V(S_i)$$

$V(\beta) = P(E)$ [the value of what would be better if E turned out to be true] + $P(-E')$ [the value of what would be better if E turned out to be false]

$$= P(E)V_E(Y) + P(-E)V_{-E}(X)$$
$$= P(E)\Sigma P(S_i/YE)V(S_i) + P(-E)\Sigma P(S_i/X(-E))V(S_i)$$
$$\therefore \quad V(\beta) - V(\alpha) = P(E)\Sigma[P(S_i/YE) + P(S_i/XE)]V(S_i)$$
$$= P(E)[V_E(Y) - V_E(X)]$$
$$> 0$$

When the determination of a statement's truth value may affect the probabilities of the possible outcomes of our alternative acts in such a way that makes a difference to which act has the highest expected value, then it is rational to precede our decision by obtaining this further data. See work by I. J. Good (1966) for a more general treatment.

The trouble with this rationale for the acquisition of new evidence is that it is too pragmatic. It does not explain why we should think that further evidence would be desirable even when our purpose is purely scientific – when our concern is solely with the truth and we give no thought to the possibility of putting our scientific judgements to work in the calculation of which practical actions we ought to perform.

(5)   Finally, I shall propose a purely epistemological solution: a synthesis of the strategies described in (2) and (4). Let us define *the error in a probability valuation* as the difference between that value and the truth value (1 or 0) of the hypothesis in question. I want to show that the expected error in our probability judgements is minimised by the acquisition of new evidence.

Our choice is between

(γ) assess H's probability on our present evidence;

(δ) first determine the truth value of E, and then assess H's probability in the light of what is discovered.

Let

$$P(H) = x$$
$$P(H/E) = y$$
$$P(H/-E) = z$$
$$P(E) = w$$

The expected error from (γ)

$$
\begin{aligned}
&= P(H)[\text{Error if } H \text{ is true}] + P(-H)[\text{Error if } H \text{ is false}] \\
&= P(H)[1 - P(H)] + [1 - P(H)][P(H) - 0] \\
&= 2P(H)[1 - P(H)] \\
&= 2x(1 - x)
\end{aligned}
$$

The expected error from (δ)

$$
\begin{aligned}
&= P(E)\,[\text{Error which should be expected if } E \text{ turns out to be true}] \\
&\quad + P(-E)\,[\text{Error which should be expected if } E \text{ turns out to be false}] \\
&= P(E)[\text{Expected error using } P(H/E)] \\
&\quad + P(-E)[\text{Expected error using } P(H/-E)] \\
&= P(E)\{2P(H/E)[1 - P(H/E)]\} \\
&\quad + P(-E)\{2P(H/-E)[1 - P(H/-E)]\} \\
&= 2wy(1 - y) + 2(1 - w)z(1 - z)
\end{aligned}
$$

Let

$$
\begin{aligned}
D &= \frac{1}{2}(\text{Expected error from } y - \text{Expected error from } \delta) \\
&= \frac{1}{2}(2x(1 - x) - [2wy(1 - y) + 2(1 - w)z(1 - z)]) \\
&= x(1 - x) - wy(1 - y) - (1 - w)z(1 - z) \\
&= x - x^2 - wy + wy^2 - (1 - w)z + (1 - w)z^2 \\
&= [x - wy - (1 - w)z] + [wy^2 + (1 - w)z^2 - x^2]
\end{aligned}
$$

But from probability theory

$$P(H) = P(E)P(H/E) + P(-E)P(H/-E)$$

that is,

$$
\begin{aligned}
x &= w\gamma + (1-w)z \\
\therefore D &= w\gamma^2 + (1-w)z^2 - x^2 \\
&= w\gamma^2 + (1-w)z^2 - [w\gamma + (1-w)z]^2 \\
&= w(1-w)(\gamma - z)^2 \\
&> 0
\end{aligned}
$$

provided that $0 < w < 1$ and $\gamma \neq z$. Thus (δ) involves a smaller expected error (γ) in our judgement of H's probability. Consequently, the truth value of E *should* be determined: this further evidence is desirable.

7

Realism

There is no shortage of objections in the philosophical literature to the general strategy and the particular concepts which have been employed here. Some of these complaints have been aired in previous chapters and, I hope, defused. In particular, I have nothing to add to the earlier arguments, in Chapter 2, for the existence of degrees of belief and their conformity to the probability calculus. Let me simply repeat what I think are the main considerations which help to allay criticism of those ideas. (1) A respectable concept need not be operationally definable. (2) The present notion of subjective probability was designed only to provide a perspicuous representation facilitating the exposure of confusion, and is not intended to serve the needs of psychology, the history of science or any other discipline.

However, there are several perspectives opposed to my approach which have not yet been explicitly addressed, and I would like in conclusion to respond directly to some of them. I will begin with Popper, who has been perhaps the most prominent and persistent opponent of Carnap-style probabilistic inductivism. Then I will discuss the bearing of Bayesianism upon the realism/instrumentalism controversy. Thirdly, I will examine an ingenious argument due to Putnam and intended to produce dissatisfaction with probabilistic confirmation theory. Finally, I will consider the criticisms of Bayesianism which motivate Glymour's 'bootstrap' conception of evidence.

Popper

According to Popper (1972) we ought never to believe that a general explanatory theory is true, or even probable; but we may often come to know that such a theory is false – when it conflicts with our data. Rational scientific inquiry proceeds by the formulation of bold (= contentful = easily falsifiable = intrinsically improbable) conjectures, by their subjection to rigorous experimental investigation, and by the invention of new hypotheses to resolve the problems, yet preserve the merits, of those which have been refuted. Progress consists in the growth in our knowledge of which theories are false, and in the increasing corroboration (survival through severe tests) of those which are as yet unfalsified, and which might therefore be true.

Popper strongly objects to the probabilistic approach to confirmation developed by Carnap and pursued in this book. In the first place, he maintains that scientists do not aim for highly confirmed (in the sense of 'highly probable') theories; for then they would be content with the assertion of tautologies. Rather, they want deep and powerful theories which say a great deal and which are, therefore, very *im*probable. In the second place, Popper denies that inductive inference is reasonable. Insofar as a theory and its predictions are not deducible from our data, there is, he thinks, absolutely no justification for believing them. Therefore, no confirmation function can be correct which assigns substantial credibility, given our evidence, either to general theories or to the unverified expectations which they entail.

I will argue that these objections are mistaken. The first derives from a failure to recognise that content and probability may each be virtues in a scientific theory, even though it is difficult to optimise both of them at once. The second rests upon a spurious distinction which Popper draws between theoretical and practical rationality.

Suppose, for the sake of argument, that our inductive practice were characterised by a Carnapian confirmation function $c(H/E) = x$. It would by no means follow, as Popper suggests, that we would not be concerned to formulate powerful theories. Our supposition has absolutely no implications one way or the other about what sort of theories we are interested in formulating. All it does is specify for each evidential state and each hypothesis that we do formulate, the degree of belief we ought to have in the hypothesis. Thus we may quite consistently supplement our supposition with the further observation that scientists are particularly anxious to discover very deep and general theories. Now, it may well be that such theories will tend to have low probabilities, given c and given our evidence; but this does not make them uninteresting. Popper's criticism of Carnap can be sustained only by confusing the supposition that some probabilistic confirmation function captures our inductive practice, with the claim (which Carnap certainly did not make) that scientists are solely concerned with highly confirmed theories. Carnap certainly did appreciate that credibility is desirable. To that end, we do experiments whose results raise the credibility of certain hypotheses at the expense of others which become disconfirmed. But this concern is quite compatible with the recognition that even survival through many severe tests will not boost the credibility of our fundamental theories to particularly high values.

Parodying Popper's line of thought, one might conclude that mountaineers are not interested in reaching the top. For if they were, why should they attempt difficult climbs such as the north face of the Eiger? Why not stick with easy slopes like Beacon Hill? The answer, quite plainly, is that although reaching the summit and engaging a challenging route conflict with one another, they are nonetheless both desiderata, and together they form a complex objective: reaching the summit by a challenging route. Similarly, one might characterise the aim of scientific investigation as the confirmation of profound theories; and, given this aim, our conformity to a c-function is not at all incongruous.

Note, in addition, that there is no conflict between our desire to confirm deep theories – to boost their credibility – and Popper's famous methodological dictum to the effect that we should attempt to *falsify* our theories. As we have seen in Chapter 5, a theory is more strongly confirmed to the extent that the tests which it survives are relatively severe. In other words, our desire for substantial confirmation is indeed best accommodated by trying to falsify the theory – by determining the truth values of those of its consequences which seem most likely to be incorrect.

Even if our c-function assigns low probability to all general explanatory hypotheses, whatever the evidence, it may nevertheless permit high degrees of belief in some of their particular observational predictions. In this, the c-function merely codifies our inductive practice: we do think it reasonable to be very confident that the sun will rise again, within twenty-four hours.

However, Popper disagrees. His 'Humean' view is that there can be no good reason to believe a general theory – even after it is well tested – and no good reason for having any particular expectation about the future. Not surprisingly, this extreme position encounters intolerable difficulties, and in a futile effort to avoid them Popper is led to an *ad hoc* and spurious distinction between two notions of rationality.

The problem is that, although we can perhaps manage without the luxury of believing explanatory theories, we cannot avoid having views about future events. Practical decisions involve some assessment of the likely consequences of alternative courses of action (or inaction). Popper is not prepared to maintain that all practical decisions are irrational, that there is no reason to prefer one to another. Consequently, he concedes that, *in some sense*, certain expectations are reasonable. Specifically, 'we should prefer as the basis for action the best tested theory' (1972, p. 22). In other words, Popper's view about those beliefs about the future which are derived from our best theories and which are relevant to our choice of action is this: they are irrational from a theoretical point of view (not being based upon deductive reasons), yet there is a further sense of 'rational' ('entailed by our best theory') in which it *is* rational to adopt them.

But this story is simply common sense and Carnapian induction dressed up in misleading jargon. Popper's conclusion, it seems to me, differs only in terminology from the view that (1) Hume was wrong; (2) certain inductive inferences are perfectly rational; (3) these cannot be shown to be rational by deducing from known facts that their conclusions will generally be true; (4) rather, their rationality derives from the fact that good reasoning includes inductive as well as deductive inference: a belief is properly deemed 'rational' when it comes from the simplest theory which accommodates our evidence. This construal of Popper's ideas dissolves their apparent divergence from probabilistic inductivism. Thus, I have no objections to Popper's positive proposals about how science should be done. On the contrary, I believe, though there isn't the space to substantiate it here, that his many sound methodological recommendations are perfectly consonant with the Bayesian approach.

Realism versus instrumentalism

Here I would like to examine briefly the contest between realism and instrumentalism, and suggest that the probabilistic approach allows a sort of reconciliation. In a nutshell, orthodox realism (taken to be the view that current scientific theories should be accepted as true) is embarrassed by the almost invariable falsification of previous theories; whereas orthodox instrumentalism (the idea that only observation statements correspond to facts, and that theories are just devices for their systematisation) is embarrassed by the difficulty of drawing a sharp principled line between observation and theoretical claims. I will suggest that the recognition of degrees of belief provides a way out of this dilemma. Perhaps, given the sad historical record, a low level of confidence is the appropriate attitude towards even our favourite theories. This position accommodates the history of science (the 'disastrous meta-induction', as Putnam calls it), but does not require the data/theory distinction and semantic dualism of orthodox instrumentalism.

First, it is important to bear in mind the difference between metaphysical and epistemological realism – notions elaborated by Putnam (1978) which descend from Kant's distinction between transcendental and empirical realism. As I see the difference, metaphysical realism is a thesis concerning the *nature of truth*, the meaning of the word, whereas epistemological realism is a thesis about its extension – a view about *what is true*. A metaphysical realist maintains that the concept of truth is not entirely captured by Tarski's disquotation schema ('*p*' is true iff *p*), but involves in addition a primitive nonepistemic idea of *correspondence with reality*. Thus, metaphysical realism stands in opposition to various constructivist theories of truth (for example intuitionism, Peircianism), according to which the surplus meaning in 'truth' involves some concept of verifiability, and to the redundancy theory which denies the existence of surplus meaning and contends that Tarski's schema is sufficient to capture the concept. Epistemological realism, on the other hand, is the view that we should accept the fruits of scientific investigation. Typically one will be an epistemological realist about certain things (for example electrons, black holes), and not about others (for example canals on Mars, the Loch Ness monster) depending upon whether one does or does not believe in them. Thus, epistemological realism concerning Xs stands in opposition to scepticism and instrumentalism. These forms of realism are by no means bound together. A metaphysical realist may embrace Cartesian scepticism and so deny epistemological realism about the external world. Conversely, one might well think that current physical theory is perfectly *true* and yet question whether metaphysical realism provides the right account of that idea.

It seems to me (1982), though I won't argue for it here, that the concept of truth is completely captured by something like Tarski's schema, and that if someone knows that the word obeys that rule then his understanding of truth is not deficient. Thus, I subscribe to the redundancy theory and oppose both of the surplus-meaning conceptions of truth. As far as I can see, the probabilistic ideas advanced in this book have no particular bearing upon that issue. But I think the question of epistemological realism, on the other hand, is very helpfully illuminated by degrees of belief.

It is not uncommon in philosophy for a controversy to feed upon some misconception shared by all the major parties in the dispute. It may then seem that the problem is to decide which of the opposing positions is right. But, in fact, none of them are; and the correct move is to locate and dissolve the misconception which fuels the dispute and infects all the adversaries. In the case of epistemological realism versus instrumentalism, it seems to me that the dialectical energy is provided by a shared tendency to ignore gradations of belief and to presuppose that scientific theories simply are accepted or not. This crude oversimplification naturally yields just two positions concerning the right doxastic stance towards our 'well-supported' theories: realism in favour of acceptance, and instrumentalism against it. However, both positions are objectionable.

Realism comes to grief over the history of science. Few old theories are accepted today; most have turned out wrong. Consider Newtonian physics, phlogiston, the ether, conservation of mass, the wave theory of light: in time, virtually every theory seems destined to bite the dust. Would it not be quite arrogant and irrational to assume that this pattern of scientific revolution will not continue, and to think that our favourite current theories will not eventually be rightly abandoned? Thus, it would be quite wrong to embrace realism and believe that our present scientific theories are true.

But if acceptance is not the appropriate attitude, what is? Instrumentalism naturally results from the assumption that our only remaining option, given the disastrous meta-induction, is to recognise that since rational belief is clearly unattainable, it could not be the proper object of scientific inquiry. Rather, science may, and does, have no higher aspiration than the economical systematisation of data; the constructions with which this organisation is to be achieved are not designed for belief, and are not intended to correspond to underlying features of the world; these so-called theories are neither true nor false.

This view certainly avoids the arrogant presumption of realism. But it surely errs in the opposite direction. Moreover, it introduces a sharp semantic distinction between the sentences recording observed facts and the theoretical machinery devised for their systematisation and this line is notoriously difficult to draw. Consider the sequence: ant, flea, amoeba, protein, atom, electron. Clearly, any sharp line between those things accepted as real, and the fictional entities invoked for organisational purposes, will seem arbitrary and quite alien to the attitudes of scientists.

The recognition of degrees of belief permits us to preserve the elements of truth in both doctrines, while avoiding their flaws. There is no reason to deny that theoretical claims possess determinate truth values, no need to turn a blind eye to the history of science, and no need to impose an arbitrary and misleading distinction upon the body of scientific claims. There are theoretical discoveries in which we have a great deal of confidence – for example that there is a planet beyond Neptune, that the species of living things evolved through natural selection, that there are units of matter, hydrogen and oxygen, which may be combined to synthesise water. Despite the disastrous meta-induction we think it unlikely that these ideas will be overthrown. It is evident that we don't distinguish crudely between those theoretical principles we are sure of, and those we expect to be supplanted. Rather, our degree of confidence varies, depending, among other things, upon the evidence, the extent to which the principle is bound up with other things we believe, and the frequency with which principles of the same type have been overthrown in the past.

I think there are self-styled instrumentalists who do not wish to deny that theoretical claims are true or false, but simply maintain that *belief* is not the appropriate attitude towards them. Also there are self-styled realists who would concede immediately that many theories deserve a low degree of confidence. Each of these trends is healthy and they converge upon the position which I have described. Thus, it is not clear to me whether the view suggested here would better be called modulated realism or sophisticated instrumentalism. I shall leave to the reader any further consideration of this delicate matter.

Putnam

In his paper 'Degree of confirmation and inductive logic' (1962), Hilary Putnam criticises Carnap for failing to recognise that the plausibility of a hypothesis will depend upon the range of alternatives which have been proposed. However, Putnam does not simply conclude that no c-function could provide a faithful representation of our inductive practice. He is mainly concerned to argue that this feature of c-methods produces a further defect, implies a clear inadequacy given the point of inductive inference: namely, that if our practice were represented by a c-function, and if we knew what that function was, then we could formulate a hypothesis which, demonstrably, we could never justifiably accept, even if it were true and no matter how much evidence in favour of it were acquired. I shall oppose this argument; and, though I sympathise with Putnam's criticism of Carnap, I suggest that Carnap's conception may be liberalised to accommodate the evidential relevance of our horizon of alternative theories.

The basic idea of Putnam's argument is as follows. Suppose we are to survey, one by one, an infinite sequence of objects $X_1, X_2, \ldots$. We begin by seeing so many red ones in a row that our c-function tells us that it is reasonable to suppose (credibility greater than $1/2$) that the next object will be red. But it turns out not to be red. We then see sufficiently many more consecutive red objects that it again eventually becomes reasonable to predict that the next object will be red. But again, the prediction is false. Suppose this pattern of accumulating sufficiently many examples of red things to justify predicting red, followed by a nonred thing, continues forever.

The sequence which actually occurs is described by a true hypothesis which we never accept, in the following sense: there is no point, no matter how many positive instances have been observed, after which we continually and consistently base our predictions upon this hypothesis:

$$(h) \; X_k \text{ is red iff } k \neq n_1 \text{ or } n_2 \text{ or } n_{3,....}$$

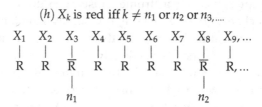

At every X_k, where $k = n_1$ or $n_2, \ldots$, our prediction will have diverged from what would be dictated by h, and all those predictions will be mistaken. It is important that we can actually calculate, using our c-function, the numbers $n_1, n_2, \ldots$ For then we can formulate the hypothesis as

$$(h) \; X_k \text{ is red iff } P(k)$$

where P is an arithmetical property. This means that h can be formulated and proposed and true, but never accepted.

So the argument goes. However, it involves a presupposition which, when made explicit, shows that the conclusion is somewhat overstated. It is assumed that the objects are observed in a specific order, namely X_1 first, then X_2 and so on. If, in fact, they are observed in a different order, then we have been given no reason to doubt that h would eventually be accepted. Thus, Putnam's argument does not establish that conformity with c may preclude discovery of the truth. It has a weaker result, that if evidence of a certain kind accumulates in a specific order then h's discovery is precluded. But this is not an obvious defect. Moreover, to block our objection it will not suffice to let X_k mean 'the kth object to be observed'. It would still remain to be shown that there is no possible evidence beyond the accumulation of instances which would lead to the discovery of h.

Admittedly, one might feel that a really superb inductive method should guarantee the eventual acceptance of a true proposal, given the sort of evidential input which Putnam assumes. However, in the first place, it has not been shown that *our* inductive practice possesses this feature; so the argument does not suggest that *we* do not subscribe to a c-function. And, in the second place, nor can we conclude that any Carnapian inductive practice can be improved upon. Grant that the eventual acceptability of any true proposal (given the stream of information assumed in the argument) is a desideratum. Nevertheless, it may well be that certain combinations of desiderata conflict with one another. And in that case it may be that a Carnapian practice, while not 'perfectly adequate', is still the best we could hope for. Putnam does discuss examples of inductive methods which do satisfy his discoverability desideratum. But, on other scores, these methods seem patently inferior to the Carnapian methods; they do not provide for degrees of credibility and do not recognise what is and should be our practice – that, given the evidence and various alternative hypotheses, the reasonable policy could be a certain distribution of belief among the alternatives.

Whatever the implications of Putnam's diagonal argument, there remains his important observation that the credibility of a hypothesis depends upon the theoretical context and the range of alternative hypotheses under consideration. Apparently, Carnap thought that the observational data alone were evidentially relevant; the particular c-functions which he devised, certainly reflect this idea. However, it is not obvious that Putnam's insight cannot be reconciled with probabilistic confirmation theory. As before, confirmation functions would prescribe degrees of belief, given a range of evidential circumstances. But these circumstances could perhaps include, not only information concerning experimental results, but also a specification of the theoretical background and the horizon of alternative proposals. In the context of Putnam's example, such a practice would require different expectations concerning the next object, depending upon whether or not h has been suggested. Given this sort of probabilistic confirmation function, it could no longer be argued diagonally that there are formulated and true, yet undiscoverable, hypotheses.

Glymour

The singular feature of Clark Glymour's (1980) 'bootstrap' account of evidence, that which primarily distinguishes it from both Bayesianism and the hypothetico-deductive (henceforth, H-D) method, is that a whole theory is not necessarily confirmed just because one of its parts is: neither evidence for p, nor even the discovery of p, is sufficient to confirm the conjunction $p \wedge q$. This is the intended result of his requirement that confirmation of a hypothesis involve derivations of instances of every quantity in it. For example, a set of observations might support

$$y = x^2 + z$$

but not the conjunction

$$y = x^2 + z \quad \text{and} \quad x = s \cdot t$$

and, therefore, not

$$y = (s \cdot t)^2 + z$$

because values of s and of t may not be calculable from the data.

Therefore, our assessment of Glymour's proposal boils down to whether we think this special feature is an advantage. Is it, in fact, intuitively plausible to maintain that the discovery of p provides no evidence for $p \wedge q$? Would this permit the accommodation of intuitive relevance-judgements which are inexplicable by means of hypothetico-deductive principles? Are accounts of various other aspects of scientific methodology facilitated by this characteristic of the bootstrap approach?

Perhaps the strongest thing to be said in favour of the opposite view, that p does confirm $p \wedge q$, is that it fits the principle: a theory is confirmed when the data ought to enhance our confidence in it. No-one is prepared to deny, I suppose, that the discovery of p should usually diminish some of the uncertainty attaching to $p \wedge q$ and ought, therefore, to augment its credibility. Consequently, if we adopt Glymour's theory, we are com-pelled to recognise that data may boost the credibility of a theory without confirming it. This result constitutes a two-pronged objection to the boot-strap approach. First, the view we have been compelled to abandon – the correlation between confirmation and enhanced credibility – has great virtues of simplicity and, at least, rough correspondence with scientific practice. Second, its rejection creates an urgent need for some alternative account of the relationship between confirmation and rational belief. Unless this is provided we are left with an enfeebled account of confirm-ation, incapable of fulfilling the very role in epistemology for which a theory of evidence was desired.

But maybe these difficulties are just the price we have to pay for a decent account of evidential relevance? One of Glymour's main com-plaints about the H-D method, to be rectified by means of the bootstrap condition, is that it fosters a misleading and unnecessary holism. When a theory T containing many hypotheses, accurately accommodates a set of data E, then, according to the H-D method, the *whole* theory is confirmed. But this should not be, says Glymour, for the theory may well contain superfluous parts which deserve no credit at all.

As it stands, this objection does not succeed. There are indeed times when T is confirmed, even though the evidence is irrelevant to some part

T' of T. But there is nothing in either Bayesianism or the H-D method to suggest that T' should be given any credit. Unless we were, mistakenly, to endorse the consequence condition (criticised in Chapter 3), we would not conclude from the fact that T entails T', that T' is supported by whatever confirms T. Thus, the views which Glymour attacks are not guilty of such counter-intuitively liberal attributions of confirmation. If anything, the danger lies in the opposite direction: namely, that these views are too strict, failing to sanction many legitimate cases of confirmation – cases, such as particular predictions of future events on the basis of past experience, where the data are not entailed by what is confirmed. So there is considerable merit in Glymour's allegation that their views do not adequately distribute praise and blame, in light of the evidence, among the various components or consequences of a theory. However, it seems unlikely that the bootstrap account can provide a remedy: its confirmation conditions are even more strict than those of the H-D method; moreover, like the H-D method, it too cannot be liberalised by adding the consequence condition without falling into absurdities. But if these deficiencies of Bayesianism and of the H-D method are not mitigated by a bootstrap requirement, there would appear to be no reason to insist upon it; to thereby abandon the principle 'p confirms $p \wedge q$', and to swallow the difficulties which this would entail.

Conclusion

I would recommend Glymour's *Why I am not a Bayesian* (1980, chapter 3), as a lucid and thorough account of all the allegedly problematic elements in our strategy. Most of the objections which he raises there have been treated in earlier chapters. Doubts about the existence and coherence of rational degrees of belief are assuaged in Chapter 2. Glymour's problem of old evidence is discussed at the beginning of Chapter 3. And his scepticism about the possibility of any probabilistic explication of 'degree of confirmation' is handled, also in Chapter 3, through the distinction between that level of support *attained* by a hypothesis and the level of support *specifically contributed* by some item of evidence. A remaining complaint is one I cannot resist quoting:

> There is very little Bayesian literature about the hodgepodge of claims and notions that are usually canonized as scientific method; very little seems to be said about what makes a hypothesis *ad hoc*, about what makes one body of evidence more various than another body of

evidence, and why we should prefer a variety of evidence, about why, in some circumstances, we should prefer simpler theories, and what it is that we are preferring when we do. And so on. There is nothing of this in Carnap, and more recent, and more personalist, statements of the Bayesian position are almost as disappointing [p. 75].

I hope that this charge can no longer be sustained. On the contrary, I think it is remarkable the enormous range of long standing problems in the philosophy of science which are illuminated by subjective probability. These include not only all the issues just mentioned, but also many more: the paradox of confirmation, the nature of surprise, confirmation of statistical hypotheses, the supposed special evidential value of prediction over and above the accommodation of data, the rationale for acquisition of new information, and the realism/instrumentalism controversy.

As I said at the beginning of this essay, my primary aim has not been to suggest a theory of the scientific method, whatever that may be. Rather, I have attempted to describe and employ a perspicuous representation for the treatment of these puzzles and questions. It has been shown that their investigation benefits tremendously: first, from a proper appreciation of the fact that belief is not an all-or-nothing matter but is susceptible to varying degrees of intensity; and second, from a precise explication of that concept, subjective probability, having the fruitful property that rational degrees of belief must conform to the probability calculus. There are, no doubt, significant differences between this invented, idealised, precisely quantified notion of subjective probability and our ordinary concepts of tentative belief and degree of conviction. And one or another of these differences are probably sufficiently important to diminish the usefulness of subjective probability in other fields, such as psychology or the history of science. Similarly, some of my explications of methodological ideas, such as 'confirmation' and 'varied data', may be quite useless in other areas of inquiry. But these limitations are not weaknesses. For the main goal of our analyses is the identification and elimination of confusion; and this is most readily accomplished if the explications are true enough to register our mistakes, yet simple enough to expose them sharply.

Bibliography

Alexander, H. G. (1958). The paradoxes of confirmation, *British Journal for the Philosophy of Science*, 9, 227–33.

Ayer, A. J. (1972). *Probability and evidence*. Columbia University Press.

Black, M. (1958). Self-supporting inductive arguments, *Journal of Philosophy*, 55, 718–25.

Carnap, R. (1950). *Logical foundations of probability*. University of Chicago Press.

Carnap, R. & Jeffrey, R. (eds.) (1971). *Studies in inductive logic and probability*. University of California Press.

De Finetti, B. (1937). On the partial equivalence condition. Translated in *Studies in inductive logic and probability*, vol. 2, R. Jeffrey (ed.). University of California Press, 1980.

De Finetti, B. (1937). Foresight: its logical laws, its subjective sources. Translated in *Studies in subjective probability*, H. E. Kyburg & H. E. Smokler (eds.). John Wiley, 1964.

Friedman, M. (1978). Truth and confirmation, *Journal of Philosophy*, 7, 361–82.

Giere, R. N. (1973). Objective single case probabilities and the foundations of statistics. In, Logic, methodology and philosophy of science, vol. 4, *Proceedings of the 1971 International Congress, Bucharest*, P. Suppes, L. Henkin, A. Joja & C. R. Morrill (eds.). North-Holland.

Glymour, C. (1980). *Theory and evidence*. Princeton University Press.

Good, I. J. (1960). The paradoxes of confirmation, *British Journal for the Philosophy of Science*, 11, 145–9; 12, 63–4.

Good, I. J. (1966). On the principle of total evidence, *British Journal for the Philosophy of Science*, 17, 319–21.

Goodman, N. (1955). *Fact, fiction and forecast*, 3rd edn. Bobbs Merrill, 1975.

Hacking, I. (1965). *Logic of statistical inference*. Cambridge University Press.

Hacking, I. (1975). *The emergence of probability*. Cambridge University Press.

Hempel, C. G. (1945). Studies in the logic of confirmation. Reprinted in *Aspects of scientific explanation*. Free Press, 1965.

Hempel, C. G. (1966). *Philosophy of natural science*. Prentice-Hall.

Horwich, P. G. (1982). Three forms of realism, *Synthese*.

Hosiasson-Lindenbaum, J. (1940). On confirmation, *Journal of Symbolic Logic*, 5, 133–68.

Hosiasson-Lindenbaum, J. (1931). Why do we prefer probabilities relative to many data?, *Mind*, 40, 23–36.

Jeffrey, R. C. (1982). De Finetti's probabilism.

Keynes, J. M. (1921). *Treatise on probability*. Macmillan.

Kuhn, T. S. (1962). *The structure of scientific revolutions*. University of Chicago Press.

Kyburg, H. E. & Smokler, H. E. (eds.) (1964). *Studies in subjective probability*. John Wiley.

Kyburg, H. E. (1970). *Probability and inductive logic*. Macmillan.

Lewis, D. (1971). Immodest inductive methods, *Philosophy of Science*, 38, 54–63.

Lewis, D. (1973). Spielman and Lewis on inductive immodesty, *Philosophy of Science*, 40, 84–5.

Mackie, J. L. (1963). The paradox of confirmation, *British Journal for the Philosophy of Science*, 13, 265–77.

Mellor, D. H. (1971). *The matter of chance*. Cambridge University Press.

Popper, K. R. (1959). *The logic of scientific discovery*. Hutchinson.

Popper, K. R. (1962). *Conjectures and refutations*. Basic Books.

Popper, K. R. (1972). *Objective knowledge*. Oxford University Press.

Putnam, H. (1962). Degree of confirmation and inductive logic. Reprinted in *Mathematics, matter and method, philosophical papers*, vol. 1. Cambridge University Press, 1975.

Putnam, H. (1978). *Meaning and the moral sciences*. Routledge & Kegan Paul.

Ramsey, F. P. (1926). Truth and probability. Included in *Foundations*, essay by F. P. Ramsey, D. H. Mellor (ed.). Routledge & Kegan Paul, 1978.

Rosenkrantz, R. (1977). *Inference, method and decision*. Reidel.

Russell, B. (1912). *Problems of philosophy*.

Skyrms, B. (1975). *Choice and chance*. Dickinson.

Spielman, S. (1972). Lewis on immodest inductive methods, *Philosophy of Science*, 39, 375–7.

Strawson, P. F. (1952). *Introduction to logical theory*. John Wiley.

Suppes, P. (1966). A Bayesian approach to the paradoxes of confirmation. In *Aspects of inductive logic*, 199–207, J. Hintikka & P. Suppes (eds.). North-Holland.

Swinburne, R. G. (1971). The paradoxes of confirmation - a survey, *American Philosophical Quarterly*, 8, 318–29.

von Neumann, J. & Morgenstern, O. (1944). *Theory of games and economic behavior*, 1st edn. Princeton University Press. (2nd edn. John Wiley, 1964.)

Index